CARTER CARBURETO

BY DAVE EM

CARTER CARBURETORS

EDITED BY
LARRY SCHREIB

TECHNICAL EDITING BY
LARRY ATHERTON

PRODUCTION BY
LARRY ATHERTON

COVER ILLUSTRATION BY
BOB ROBE

ISBN 0-931472-11-3
PART No. 11-3

S-A DESIGN BOOKS, 515 WEST LAMBERT, BLDG E, BREA, CA 92621-3991

TABLE OF CONTENTS

CHAPTER 1

IN THE BEGINNING

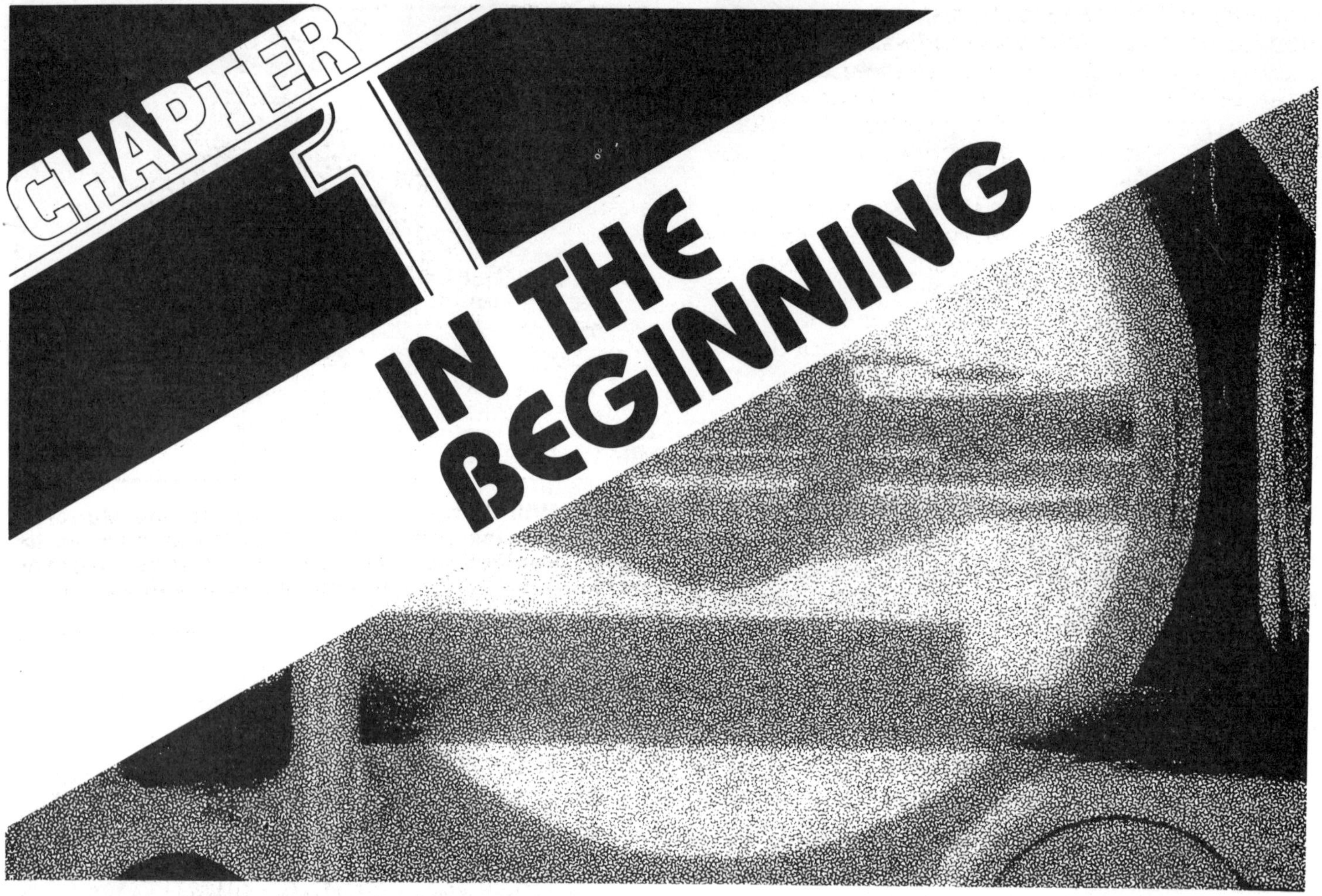

The founder of Carter Carburetor Company—Mr. Will Carter. Like most inventors of the early 1900's, he wasn't very adept at handling business matters and was forced to relinquished control of the company seven years after it was founded.

THE CARTER CARBURETOR COMPANY

As the dawn of the age of mechanized transport grew into a frenzy of mid-day activity, youthful inventors imbued with native mechanical genius gravitated toward the emerging automobile business. And in the first years of this century, names that have since become immortalized in the annals of automotive history made their first appearance before the American public: David Buick, Ransom Olds, Henry Ford, Harry Stutz, Fred Duesenberg. By 1910 these geniuses of engineering and business brought dramatic change to the face of automobile manufacturing.

But there were other lesser known inventors whose contributions were of equal importance. However, since their endeavors were directed not toward building complete automobiles, but to the creation and refinement of products required by the auto manufacturers, history has not treated their accomplishments quite so reverently. Will Carter was one of these men. Born in 1884, just outside Union City, Tennessee, Carter received only five years of formal education. But at the age of 17, demonstrating a flair for things mechanical, he opened a repair shop servicing bicycles, guns and virtually any other type of mechanical apparatus that came through his front door.

In 1902, Carter felt the limitations of his rural location, and since St. Louis, Missouri, was the only large city within reasonable distance to his small shop, he moved on to the opportunities presented by an urban environment. At best these were limited as the automobile population of St. Louis was limited—less than a dozen cars. But Carter offered his talents and began expanding the base of his repair business.

As was the case with most inventors of the day, Will Carter was a tinkerer. As if poor quality gasoline wasn't enough for early motorists to contend with, the devices used to meter that gasoline provided an additional source of irritation, so Carter began to experiment with techniques and methods of improving carburetor operation. His designs, translated into wooden models, served as the cores for sand molds from which an improved, cast bronze carburetor would appear.

FROM TINKERER TO MANUFACTURER

Carter's new device brought greater accuracy to the process of metering fuel and mixing it with air, and as word of its superiority spread, demand rose to a sufficient level that in 1909, with the financial backing of a friend, Will Carter founded the Carter Carburetor Company. The following year, the Model C carburetor was patented. It was an updraft design and incorporated an air valve. The Model C was advertised as offering "dignified acceleration," and other literature of the era stated that the carburetor "has conclusively proved the established principle of automatic-multiple jets. It has separate adjustments for low, intermediate, and high speeds, however its action is entirely automatic and these adjustments, when properly made, are fixed, requiring no further attention."

By 1911, Carter had designed and built the first downdraft carburetor. It was augmented with a unique fuel handling system which utilized manifold vacuum to pump fuel from the main gas tank to a small reservoir located above the carburetor. The pump assembly used a diaphragm constructed of

One of Will Carter's earliest creations, the Model C carburetor was patented in 1910. It was claimed to offer "dignified acceleration." The Dodge Brothers were among the first automakers to use this carb.

It might be said that the move into Carter's first factory, in 1915, was a "shoe in." The building was formerly occupied by a shoe manufacturer.

Carter went to war with this carburetor—the M-2. It served as original equipment on Liberty trucks used by the U.S. Army during World War I.

Carter's model "FO," a simple but reliable design, was used on a variety of vehicles between 1910 and 1920.

linseed-treated raincoat material.

But alas, the inability to properly discharge financial matters seems to be an adjunct to the inventor psyche and by 1916, the Carter Carburetor Company was on the skids. It was reorganized in that year and Will Carter was left without a management role. Six years later the company was purchased by the American Car and Foundry Company, now known as ACF Industries.

Until 1925 Carter produced replacement carburetors, but with its first original equipment order from a major manufacturer (Chevrolet), the direction of the company began to change. Chrysler began purchasing carburetors in 1928 and throughout the 1930s many other manufacturers—some of whom have since joined their ancestors in the great wrecking yard in the sky—became original equipment customers: Nash, Hupmobile, Willys, Ford and later Buick, Oldsmobile and Pontiac.

THE FOUR-BARREL ERA

1952 was an important year in the development

The Beach Boys made it immortal—"my four-speed, dual-quad Positraction 409." The dual quads were Carter AFBs and a performance mystique has surrounded the carbs ever since this famous installation.

Brothers or identical twins? Many people mistake the AVS for an AFB. While the two models are very similar, the AVS, introduced in 1966, utilizes a spring-loaded—rather than counterweighted—secondary air valve.

The first American-built four-barrel was produced by Carter in 1952. It was first used by Buick on a straight-8 engine.

of carburetion for two major reasons. First, there was the debut of the BBD two-barrel on a DeSoto. This efficient carburetor, albeit with significant changes, is still being produced. (Unfortunately the same can not be said of the venerable DeSoto which Chrysler retired many years ago.) The second, and more important event—one that inexorably changed the history of the carburetor—was the introduction of the world's first four-barrel. This original design, called the WCFB (for Will Carter Four Barrel), appeared atop the aging, Buick straight-eight engine.

Compared to later four-barrel designs, the WCFB seems indeed archaic; weighing 18 pounds and comprised of a cast iron base, a zinc main body, and an aluminum air-horn assembly. But this original four-barrel carburetor broke new ground because it was widely used on high-performance engines, and still remains an excellent carburetor for small displacement V-8 engines. Although the WCFB remained in production through the mid-Sixties, the AFB, a more capable design introduced in 1957, stole most of its thunder. The AFB became "the" choice for performance engines of the early Muscle Car era. Finally, after nearly ten years of dominance, the AFB was largely supplanted by the AVS (Air Valve Secondary) in 1966 and later original-equipment applications.

As emissions regulations grew more stringent, new priorities in the field of automotive carburetion were mandated, and in 1971 Carter introduced the Thermo-Quad as an original equipment carburetor on 340 cubic inch Chrysler engines. The TQ (Thermo-Quad) was the first production carburetor to use a plastic (phenolic resin) main body, but it has been used only by Chrysler Corporation. However, Carter markets the TQ as a replacement for the Rochester Quadra-Jet, so it is commonly installed on Chevrolet engines.

However, the TQ is not the whole story of modern high-performance carburetion from Carter. Because the AFB was very successful atop such immortal engines as the 409 Chevrolet, 389 Pontiac (specifically those installed in GTO'S) and Chrysler Hemi, the mystical aura surrounding the AFB continued long after the carburetor was phased out. The long-standing reputation created demand, and in the mid-1970's, Carter reintroduced the AFB as the Model 9000. Updated with an electric choke, emissions device connections, original equipment throttle linkage compatibility, and recalibrated to meet emissions requirements, the 9000 Series was designed as a bolt-on replacement for installation on Chevrolet, Chrysler and Ford engines. Although most popular in performance applications, these carburetors are also excellent at squeezing increased fuel economy from late model V-8s.

As one of only two independent carburetor manufacturers in this country, Carter continually experiments with products that are in concert with prevailing priorities. But as four and six cylinder engines achieve greater popularity (particularly the V-6 engine—see V-6 PERFORMANCE, available from S-A Design Publishing Co., page 128), development of aftermarket intake manifolds that will mate the TQ and 9000 carburetors to the new generation of down-sized engines will surely result. So whether you drive a V-8, a V-6, or a four, there just may be a Carter carburetor in your future.

The currently available 9000 Series is an updated model of the original AFB which was unveiled in 1957. The new version incorporates changes that make it compatible with current engine manifolding, ignition, etc.

It's easy for an engine to keep its cool when it's equipped with a Thermo-Quad. The main body is made of phenolic resin which serves to keep the fuel in the float chamber cool and dense for improved power output. Introduced in 1972, the popular Thermo-Quad has been used as both an original equipment and replacement carburetor.

Chrysler has used Carter carburetors for many years, and is the only auto manufacturer to use the Thermo-Quad in original equipment applications. This is a 340 cubic-inch smallblock.

Although the Chrysler Hemis were offered with a variety of carburetors, many race and street applications—including the infamous Street Hemi were equipped with dual AFBs. The AFB is known for its powerful starting-line "launching ability."

CHAPTER 2 CARBURETOR TYPES

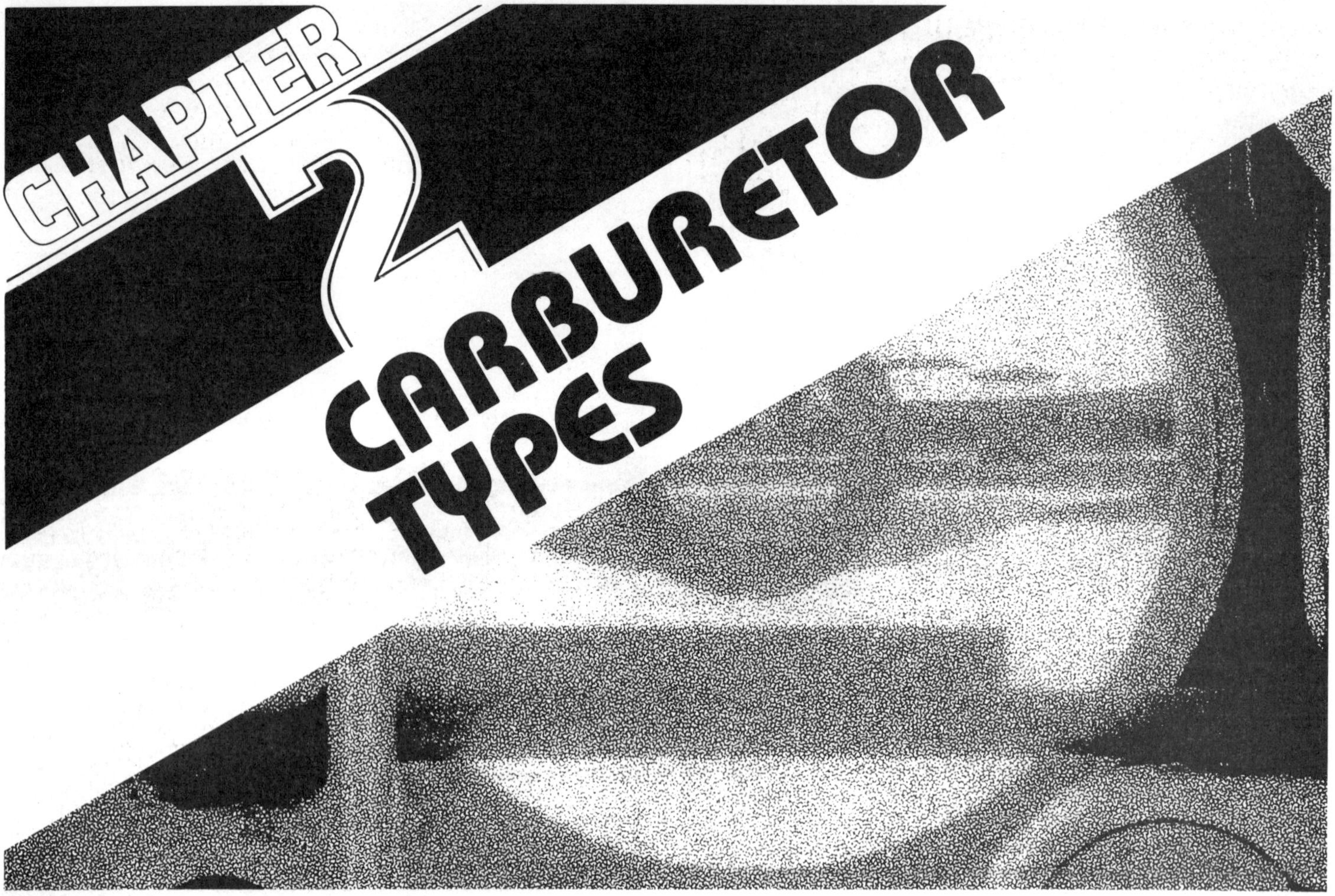

Although the WCFB hasn't been produced for a number of years, it still enjoys a degree of popularity, especially with restorers of early Chevy V-8 engines. This factory set-up is installed on an immaculate 1957 Nomad belonging to Eddie Farr of Houston, Texas.

Some people still take their WCFB's to the races. This Corvette is competing at the historic auto races at Laguna Seca Raceway. Early dual-quad Corvettes can also be found at autocross competitions organized by local Corvette clubs.

BASIC CARTER MODELS

Over the years, Carter has produced four distinct models of four-barrel carburetors—the WCFB, AFB, AVS and Thermo-Quad. The company has also produced a number of two-barrel designs including the WCD, WGD, and BBD. Aside from some main jets and needle-and-seat assemblies, very few parts are interchangeable. Each model utilizes parts of unique design to prevent cases of mistaken identity.

THE WCFB

The WCFB, introduced in 1952, flowed approximately 385cfm and weighed close to 18 pounds. It was comprised of three basic castings: a cast iron throttle body, a zinc main body, and an aluminum air horn. The WCFB may be easily identified by its two "rectangles within a circle" air horn, and a throttle

In addition to displacing 392 cubic inches, the early Chrysler Hemi was physically immense and extremely heavy. It takes a lot of carburetion to feed such a monster but these dual AFB's are up to the task.

Although the AFB was introduced in 1957, it wasn't until 1962 that it established a reputation as a high-performance carburetor. Using two inline AFBs, the Chevy 409—with a rated 409 horsepower—put the Carter aluminum four-barrel in the spotlight.

With engines like this dual four-barrel equipped 426 cubic inch Ramcharger from the early 60's, Chrysler didn't hurt the high-performance image of the AFB a bit.

body which invariably takes on a rust color once it has seen front-line action in an engine compartment.

The WCFB has not been produced for some time, so it is almost exclusively found in older factory installations. From 1956 until 1961, Chevrolet utilized a dual-WCFB carburetion system on the 265 and 283 cubic inch engines. Many of these were installed in Corvettes, so there is considerable interest in WCFB amongst members of the Corvette restoration fraternity. Single WCFB installations were used up to 1965 on some Chevrolet engines, on 1955-57 on Chryslers and DeSotos, on 1952-55 Oldsmobiles, and on 1955-60 Studebakers.

The WCFB, given its relatively low air flow capacity is best suited for small V-8s. While it is a good carburetor with straightforward metering circuits and counter-weighted, air-valve secondaries, a later model is preferable since new replacement parts can be difficult to locate. With the exception of rebuild kits, WCFB components will have to be ferreted out of a wrecking yard.

THE ORIGINAL AFB

By 1957 engine sizes had increased to the point where a larger four-barrel carburetor was necessary. To meet this need, Carter introduced a completely new, lightweight design, called the AFB (for Aluminum Four Barrel). Where the WCFB made use of three major castings, the AFB utilized only two, and both were made of aluminum. The AFB is the quintessential Carter four-barrel, and has been offered as original equipment by every major U.S. automaker. It was the carburetor that virtually launched the early "Muscle Car" era, and it put Carter in the business of high performance.

Specific air flow ratings for early AFBs are not readily available since Carter did not flow rate their products in those days. However, judging by venturi and throttle bore diameters, it appears that the flow range is between 450 and 625cfm, with variations dependent upon the original application. During the early 60s, it was common knowledge that the model used on 425 cubic-inch Buick Wildcats was larger than these used on dual four-barrel equipped 409 Chevrolets. However, the Chevrolet unit (which flowed 625cfm) was quite popular and was sold for a number of years as a replacement carburetor.

THE LATE-MODEL AFB

More recent versions of the AFB are better developed and more suitable for use on modern engines. Carter offers both general performance (9000 series) and racing (Competition Series) lines. The 9000 series carbs are designed for easy installation, and there are specific models calibrated for GM, Ford or Chrysler engines. These AFBs also feature electric chokes, simple linkage connections, ports for vacuum advance and PCV valve activation,

The 9000 Series AFBs are available with air flow ratings of 400, 500 or 625cfm. Additionally, all models are built to be compatible with a particular group of engines. This carburetor, a model 9625, is easily installed on the 350 cubic-inch Chevy smallblock.

Chrysler engines typically utilize their own style of throttle and transmission kick-down linkage. With that in mind, Carter designed the model 9626 AFB to fit both smallblock and big-block Mopars.

and they are available with flow capacities of 400, 500 and 625cfm. Competition Series AFBs are offered in 500, 625 and 750cfm capacities with either a manual choke or none at all.

All current AFB models contain dual-pattern drillings on the mounting flange, so the carbs will not only fit manifolds of the standard bolt pattern (5-1/8- by 5-5/8-inch, originally designed for carbs such as the Holley 4150/4160 and Ford/Autolite 4100), but also for spread-bore models (such as the Rochester Quadra-Jet, Holley 4165/4175, and Carter Thermo-Quad which have a 4-1/4- by 5-5/8-inch bolt pattern). However, it should be noted that bolt pattern compatibility does not assure proper fit. Due to variations in manifold design, there exists a possibility of interference between manifold and carburetor appendages (insufficient room for the linkage to rotate through its full travel, etc.) Although not very common, such incompatibilities should be considered.

Selection of the appropriate 9000 Series AFB is relatively simple since there are only 11 part numbers. Three (9400, 9500, 9625) are designed and calibrated for Chevrolet engines and differ in flow capacity; two (9501 and 9626) are for Chrysler applications and one (9627) for Ford.

The remaining five part numbers are essentially carbon copies of the preceding six, but have provisions for EGR connections; two numbers pertain to Chevrolet applications (9635 and 9510), one to Ford (9637) and two to Chrysler (9636 and 9511).

In Carter's 9000 scheme, part numbers directly relate to flow ratings; a 9400 flows 400cfm, a 9500 is a 500cfm carburetor and a 9625 will flow 625 cubic feet of air per minute. Part numbers that end in the exact flow rating (i.e. zero or five) are designed for Chevrolet applications; adding 1 indicates Chrysler compatibility (for example, part number 9626 is a 625cfm unit for Chrysler engines) and the addition of 2 refers to units designed for Ford (9627). Adding 10 to any of the basic part numbers indicates EGR provisions. The performance line also includes a special carburetor available as part 3721. It is a 575cfm carb originally used on the high-performance 1964-1965 Corvette 327 cubic-inch engine. In addition, a complete line of original-equipment replacement carburetors is available from Carter.

The fact that the 9000 Series carbs are listed as performance replacements can be somewhat misleading. While these AFBs are well suited to high-performance engines, they are also ideal for applications where economy is of prime importance. The term "performance" simply refers to products that are other than original equipment replacements; in this instance, the old "performance equals more horsepower" equation may not hold true. As an example, installation of a 400cfm AFB on an engine originally equipped with a larger carburetor would *reduce* horsepower. However, the smaller carb offers a potential for *increased* fuel economy, and considering its relatively low price, it is often

In addition to working well on stock engines, AFBs can also be used successfully on modified powerplants such as this 350 smallblock installed in a Camaro. The carburetor is complemented by a high performance cam and Edelbrock Performer dual-plane intake manifold.

choosen specifically for this application.

All of the 9000 Series AFBs contain 1-7/16-inch primary throttle bores and with the exception of the 400cfm model, 1-11/16-inch secondary throttle bores. (The 400cfm version has 1-7/16-inch secondary bores.) Venturi diameters range from 1-1/8- to 1-3/16-inch on the primary side and from 1-1/4- to 1-9/16-inch on the secondaries, as illustrated in the accompanying chart. Competition Series AFBs are similarly sized as indicated in the chart by the CS (Competition Series) part numbers.

CARB. NO.	CFM	VENTURI DIAMETER		BORE DIAMETER	
		PRI.	SEC.	PRI.	SEC.
AFB 3721	575	1-1/4	1-9/16	1-9/16	1-11/16
AFB 9400	400	1-1/8	1-1/4	1-7/16	1-7/16
AFB 9500	500	1-3/16	1-1/4	1-7/16	1-11/16
AFB 9501	500	1-3/16	1-1/4	1-7/16	1-11/16
AFB 9510	500	1-3/16	1-1/4	1-7/16	1-11/16
AFB 9511	500	1-3/16	1-1/4	1-7/16	1-11/16
AFB 9625	625	1-3/16	1-9/16	1-7/16	1-11/16
AFB 9626	625	1-3/16	1-9/16	1-7/16	1-11/16
AFB 9627	625	1-3/16	1-9/16	1-7/16	1-11/16
AFB 9635	625	1-3/16	1-9/16	1-7/16	1-11/16
AFB 9636	625	1-3/16	1-9/16	1-7/16	1-11/16
AFB 9637	625	1-3/16	1-9/16	1-7/16	1-11/16
CS 4758	500	1-3/16	1-1/4	1-7/16	1-11/16
CS 4759	625	1-3/16	1-9/16	1-7/16	1-11/16
CS 4760	750	1-7/16	1-9/16	1-11/16	1-11/16
CS 4761	500	1-3/16	1-1/4	1-7/16	1-11/16
CS 4762	750	1-7/16	1-9/16	1-11/16	1-11/16

The AVS can be found on a number of mid-to-late Sixties Chrysler engines, however, since it looks so similar, it is often confused with the AFB. This 273 Chrysler smallblock was an excellent street performer, with low-end torque boosted by the low-speed throttle response of the AVS.

THE AVS

At first glance, an AVS (Air-Valve Secondary) can easily be mistaken for an AFB. While the two models are quite similar, there are significant differences. Primary among these are the AVS's spring-loaded air valve at the very top of the secondary air horn, and the lack of secondary booster venturis. By comparison, the AFB utilizes a counter-weighted air valve just above the secondary throttle plates and the usual booster venturis.

The AVS, first used to a limited degree in 1966 on certain Chevrolet models, never achieved widespread popularity, largely because most people thought it was an AFB. In fact, the AVS can offer greater performance potential than its more famous

Although it is known primarily as an original equipment Chrysler carb, the AVS was also used on some Chevrolet engines beginning in 1966. Note the secondary air valve; it looks very much like a second choke.

The best way to see the differences in the AFB and AVS design is to place the two carbs side-by-side (the AFB is on the left). A spring-loaded secondary air valve at the top of the air horn is the AVSs most distinguishing characteristic, but also note that it has dimpled rather than flat metering rod chamber covers (arrows) denoting three-step rather than two-step metering rods.

counterpart because the opening rate of the secondary air valve is easily adjustable. However, the AVS is available only in standard replacement configurations, so matching throttle and choke linkage to late-model engines can be a problem. This is complicated by the fact that most original AVS applications were on 1968-1971 Chrysler products, which used throttle cables and unique (in-manifold) choke bi-metal spring.

The Thermo-Quad is an emissions-type four-barrel with extremely small primary and large secondary throttle bores. It's concept is similar to the Rochester Quadrajet, although functionally they are quite different.

THE THERMO-QUAD

Introduced in 1971 as an original-equipment carburetor on 340-cubic inch Chrysler engines, the Thermo-Quad is a spread-bore type carburetor with small primary and large secondary throttle bores. Like the AVS, the Thermo-Quad utilizes an adjustable secondary air valve, but unique to the TQ is its main body which is manufactured of a black phenolic resin. This type of construction serves to reduce fuel bowl temperatures since phenolic material has excellent temperature insulating characteristics. Lower temperatures reduce percolation and increase power since fuel is more dense when it is cooler (when fuel becomes extremely hot, it loses density, which can result in lean conditions). Chrysler is the only automaker to have used the TQ as original equipment, and has used the carb on many of its four-barrel equipped vehicles since 1972.

In addition to original equipment and standard replacement models, performance Thermo-Quads are available under the 9000 Series label. Rated at 800cfm, part numbers 9800 and 9810 are intended for Chevrolet applications, and 9801 and 9811 for Chrysler engines (part numbers with 10 added contain EGR provisions). Both versions feature 1-

This original-equipment Thermo-Quad is a dead ringer for a 9000 Series model, but only from the outside. Due to exhaust emissions considerations, this Chrysler model incorporates many minor internal changes.

As is the case with the AFB, the 9000-Series Thermo-Quad is available under a variety of part numbers, each designed and calibrated for a specific engine group. This carburetor, part number 9800, is an excellent replacement for the Quadrajet on Chevrolet smallblocks.

The "Super Quad" label identifies the carb inside as a 9000 Series model. Two groups of carbs are available, one with EGR (exhaust gas recirculation) provisions, the other without.

While Chrysler is the only automaker to have used a Thermo-Quad as an original equipment carb, it isn't the only truck builder to have done so. International Harvester shares the honor. This is the carburetor installation station on the International assembly line.

3/16-inch primary venturis, l-3/8-inch primary throttle bores and 2-1/4-inch secondary venturis and bores. 9000 Series Thermo-Quads feature an electric choke, PCV provisions and ported-spark vacuum connections. The 9000 Series Thermo-Quads differ slightly from the original equipment models but most parts are interchangeable.

Due to its small-primary/large-secondary throttle bore configuration, most Thermo-Quads are used for other than original equipment replacement, except on engines originally fitted with Rochester Quadra-Jets. However when used in such applications, the fuel line and possibly some vacuum lines must be rerouted. Another source of irritation is the TQ's single fuel inlet located on the rear portion of the air horn; the Quadra-Jet admits fuel through an inlet situated at the front of the main body.

THE BBD TWO-BARREL

Carter's model BBD two-barrel can cause quite a degree of confusion because of the wide variation in the appearance of carburetors listed under this designation. First used in 1952, the BBD has become the standard-bearer of Carter two-barrels and is still used on many Chrysler V-8s. The "BB" portion of the nomenclature stands for Ball and Ball and pertains to the basic design, which Carter purchased from the Ball brothers (whoever they were); the "D" represents "dual" barrels as contrasted with the "BBS" single-barrel carb.

BBD carburetors have been produced in a variety of sizes, the most common being 1-1/4-inch and 1-1/2-inch throttle bore versions. Although there is a marked external difference between older and newer models, internal functioning is identical. The BBD may not have all the glamour of an AFB or Thermo-Quad, but it is an excellent choice when a two-barrel carburetor is required.

Although specific fuel circuits differ between models, all Carter four-barrels—and the BBD two-barrel—function almost identically. Once a basic understanding is achieved, the carburetors prove to be easy to rebuild and modify for increased performance, fuel economy, driveability or other specific characteristics.

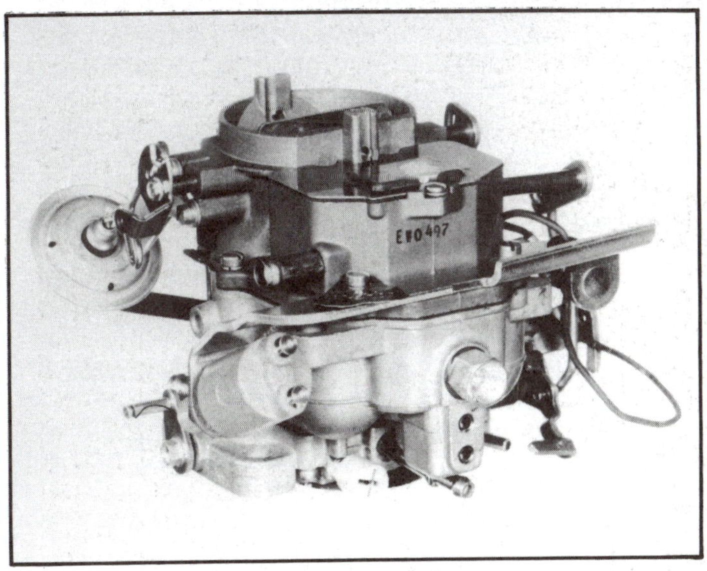

The Carter BBD two-barrel has been kicking around since 1952, but it has survived well into the 1980's. Changes in calibrations and circuitry have adapted it for "emissions" engines, including Chrysler's lean-burn 318.

CHAPTER 3 CARBURETOR SELECTION

HOW BIG?

Prior to the development of the Carter WCFB, the installation of multiple carburetors was the only effective means of improving engine breathing. Early attempts at high-performance V-8 carburetion typically consisted of an aluminum intake manifold and two, three, four, six or eight Stromberg two-barrel carburetors. Air flow ratings were not generally available, but back in the days of the small-displacement Ford flathead V-8 and Chevrolet 283 smallblock, there was little need for high flow ratings.

Later efforts at multiple two-barrel carburetion used the larger Rochester 2GC, and in the late Fifties even General Motors got into the act with triple two-barrels on the Oldsmobile "J-2," Pontiac, and Chevrolet 348. But the most unusual (although ill-fated) development in multiple two-barrel manifolds was the handiwork of a company by the name of Man-A-Fre. By positioning a Rochester two-barrel directly over each pair of cylinder head ports, Man-A-Fre marketed "direct port induction" as the state-of-the-art in manifolding. If nothing else, the four carburetors looked impressive, but for most users, performance never lived up to the enthusiastic promises of the marketeers.

When awareness of carburetor air-flow ratings became commonplace, "bigger" was typically equated with "better" (and too much was just enough), since even the largest four-barrel of the time was not capable of over-carbureting a highly

The Stromberg two-barrel was one of the first carbs used in high-performance applications. It was an original equipment carburetor on Ford flathead V-8's but spiced up the life of many modified four bangers.

This triple-Stromberg 97 installation was one of the earliest approaches in improving performance through increased air flow. Current four-barrels offer a higher flow capacity than five of the old Strombergs.

For a time rumors abounded that the Autolite inline four-barrel was to be the ultimate high-performance carburetor. While this inline concept offers a number of theoretical benefits, it never proved out in practice.

modified overhead-valve engine. The early WCFB flowed only about 385cfm. Consequently a dual-WCFB installation would flow only 770cfm. (It is interesting to note that dual-WCFB four-barrels first appeared as original equipment on the 225-horsepower 1956 Corvette however, a factory triple two-barrel installation wasn't offered until 1957 when three Rochester 2GCs were fitted on certain Oldsmobile and Pontiac engines.)

By comparison, the largest-capacity Carter four-barrel currently produced—the Thermo-Quad—

Factory-installed, dual, Carter four-barrels first appeared on the 225-hp Corvette engine of 1956. Combined air flow capacity was only 770cfm.

flows 800cfm, and 750cfm AFBs are also available. Obviously, with the general availability of such high capacity carburetors, bigger no longer necessarily equals better. In fact, a smaller four-barrel is often preferable to an extremely large one. While a 625cfm unit may not offer the glamour of dual AFBs on a ram-type manifold, the former will provide better driveability, greater fuel economy and crisper throttle response than the latter, on all but large displacement super-high-performance and race engines. A single four-barrel is also considerably lower in price and easier to maintain.

HOW TO DETERMINE FLOW REQUIREMENTS

Choosing the correct performance carburetor for a particular installation involves careful analysis. The first step is to determine the maximum air flow requirements of the engine. On the surface, this appears to require little more than converting engine size from cubic inches to swept volume (air required to "fill" the cylinders) in cubic feet, then selecting a carburetor that offers a corresponding air flow capacity. However, to arrive at an optimum selection, induction efficiency, the usable rpm range, the intended use of the engine, plus the total number of throttle bores must be taken into consideration.

AIR FLOW: RATINGS VS. REALITY

The basic mathematical formula for relating engine size and maximum RPM to carburetor airflow

The Pontiac GTO musclecar has always enjoyed an excellent performance reputation, as its engine would dish out buckets full of low-speed and mid-range torque. The base engine was equipped with a Carter AFB, properly sized to avoid over-carburetion.

Although flow capacity is an important consideration, air/fuel calibration also influences overall engine performance. This test/calibration facility accurately meters both fuel and air flow, and the complexity of this equipment illustrates why these measurements are beyond the capabilities of most enthusiasts; leave air bleed and internal modifications to the experts.

capacity is:

$$\text{Carburetor cfm} = \frac{\text{Engine CID}}{2^{*}} \times \frac{\text{Maximum RPM}}{1728^{**}}$$

***Division by 2 is required because a four-cycle engine has an intake stroke every other revolution.**

**** Factor to convert cubic inches to cubic feet; 12x12x12=1728**

For ease of use, this formula may be rearranged:

$$\text{Carburetor cfm} = \frac{\text{Engine CID} \times \text{Maximum RPM}}{3456}$$

This simplified method of calculating the required flow rating assumes 100% volumetric efficiency (i.e., no resistance to air flow in the induction system) and does not consider the pressure differential (pressure drop on the test bench that draws air through the carburetor) used in flow rating the carburetor. Of these two factors, the flow-test pressure differential will have the greatest effect on relating engine cfm requirements to carburetor cfm flow. However, since they agree on little else, most carburetor manufacturers have standardized on: 1) 3.0-in/Hg (in/Hg: inches of mercury) as the pressure drop at which two-barrel carburetors are rated, 2) 1.5-in/Hg as the pressure drop at which four-barrel carburetors are rated. Since air flow increases as the pressure drop is raised (in a non-linear fashion; doubling the pressure drop produces less then double the flow), the flow ratings of two and four-barrel carbs can be equated as follows:

$$\text{Flow at 1.5-inch/Hg} = \frac{\text{cfm rating at 3.0-inches/Hg}}{1.414}$$

$$\text{Flow at 3.0-inches/Hg} = 1.414 \text{ X (cfm rating at 1.5-inch/Hg)}$$

The differences in flow ratings (3.0- vs 1.5-in/Hg) are based on what is presumed to be the maximum vacuum in the intake manifold under load at wide-open-throttle (WOT). These pressure-drop values are intended to relate flow bench data to "real world" flow on an engine. It can be simply stated that, with any given engine, as the total throttle-bore and venturi areas increase, maximum WOT intake manifold vacuum decreases. In other words, in theory it is assumed that at WOT a conventional passenger-car engine will never realize an intake manifold vacuum higher than 1.5-in/Hg when equipped with a four-barrel, nor more than 3.0-in/Hg when fitted with a one- or two-barrel.

However, the mere fact that an engine is fitted with a two- or four-barrel carburetor does not guarantee that the engine will realize the specific pressure differential at which the carburetor was rated. For instance, installation of a very large two-barrel may drop manifold vacuum below 3.0 in/Hg at full throttle, in which case, the carb will not reach its rated flow capacity. Conversely, a small four-barrel may exceed its rated capacity if manifold vacuum

Too much of a good thing is still too much. The modest size carburetors used in lower Stock and Super Stock classes demonstrate that "big number" carburetors are not essential for impressive performance. The key is matching air flow capacity to specific engine requirements

never drops to 1.5 in/Hg at WOT. As a result, it is entirely possible that a 9400 series AFB, rated at 400cfm, will flow over 500cfm, and that the 500cfm 9500 series will achieve a capacity of over 700cfm. The fact that AFBs utilize an air valve to control flow through the secondary throttle bores—thereby regulating carburetor air flow to engine demand—increases the tendency of these small four-barrels to maintain WOT manifold vacuum above l.5 in/Hg.

Obviously, carburetor air flow ratings are not intended as absolutes, rather, they are relative guidelines. Depending on engine displacement and volumetric efficiency, air flow may exceed or fall below the rated capacity of a particular carburetor. In most instances, the manufacturer's rating will be a very usable approximation, so it really doesn't pay to become entangled in long calculations involving venturi shapes and air flow peculiarities—even if you *do* have the latest model electronic calculator. The best bet is to follow Carter's recommendations. As an example, the Carter catalog contains a chart of suggested carburetors for street, and combination street/race usage. Each of the categories assumes a specific level of engine modification (i.e., a certain degree of volumetric efficiency; the greater the state of modification, the greater the volumetric efficiency) and carburetor size is adjusted accordingly.

Moreover, there are instances where particular engine/chassis combinations are intended for other than commonly encountered applications. In these situations, calculating engine air flow requirements may be advantageous. By way of example, consider a 350 cubic inch power plant with a maximum engine speed of 6,500 RPM.

$$\frac{\textbf{350 (CID) X 6,500 (RPM)}}{\textbf{3456}} = \textbf{658 cfm}$$

Dual four-barrel installations seem to contradict the equation that matches air flow to engine size. In fact, they don't. Large displacement race engines—like this 409 Chevy—designed for high RPM power, will often produce more peak power with dual carburetors. However, low-speed power will usually suffer.

With the AFB, AVS and Thermo-Quad to choose from, it isn't difficult to find a carburetor with the proper air flow rating. The smallest AFB flows 400cfm while the large Thermo-Quad is capable of flowing 1000cfm.

The 625cfm AFB—available for Chevrolet, Ford and Chrysler engines—is an excellent choice for larger or highly-modified powerplants. This 9627 AFB is mounted on a 351 Ford.

In the early Sixties, Carter developed this experimental three-barrel AFB for the Pontiac racing effort. Flow capacity was increased by replacing the two secondary throttle plates with one large throttle valve. Very few of these carbs were produced and are highly-valued collector items.

When computing engine air flow requirements, volumetric efficiency should always be considered. In this regard, a hemispherical combustion chamber—such as used on this 1966 426 Dodge—is superior to one with a wedge shape. In theory, the Hemi can tolerate larger carburetors than a comparable engine with wedge chambers.

Volumetric efficiency is influenced by the length of the intake tract. To take advantage of "ram tuning," Chrysler released a series of engines that utilized tuned intake runners: the Ramchargers. This extraordinary engine produced in the Fifties used long manifold runners tuned to maximize engine torque.

A QUESTION OF EFFICIENCY

With the 658cfm requirement, a 625cfm carburetor would be very close to the optimum selection. Therefore, 9625 AFB would fulfill the air-flow requirements and serve nicely for applications such as road racing, street/bracket-type drag racing, slalom, gymkhana/autocross competition and off-road use. It should be noted that no adjustment has been made for volumetric efficiency, and obviously, few engines ever reach 100% v.e. Most operate in the 70%-95% v.e. range. Therefore, the 658cfm figure cited in the above example is strictly theoretical, and should be modified as follows:

VE—%	CFM requirement
70	460
75	493
80	526
85	559
90	592
95	625
100	658

Volumetric efficiency varies throughout the RPM range and is usually highest at engine speeds where maximum torque is being produced. For general reference use this "rule of thumb;" 75% v.e. for low-performance engines, with late-model smog motors slightly lower; 85% v.e. for most high-performance engines; 95% and higher v.e. is appropriate for supercharged and fully-modified racing engines. What this all means is that instead of consuming 350 cubic inches of air every two revolutions, a 350 CID engine will use:

262 cubic inches operating at 75% v.e.
297 cubic inches operating at 85% v.e.
332 cubic inches operating at 95% v.e.

Natural laws of flow dynamics govern the volumetric efficiency of any engine at a particular RPM. Intake manifold efficiency, valve and port size, camshaft timing and exhaust manifold configuration are a few of the more readily identifiable factors affecting volumetric efficiency. However, there is an additional factor inherent in all engines that helps improve v.e., in the same way a supercharger increases horsepower. Since the vacuum (pressure differential) created by a piston moving downward in a cylinder during the intake stroke is not sufficient to permit the complete filling of that cylinder (with the piston at the bottom of its travel), an additional force

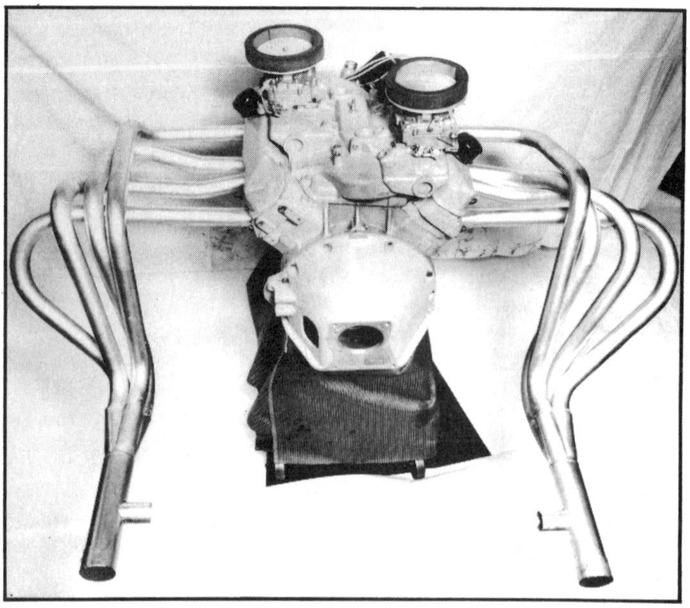

Later Ramcharger engines used offset carburetor mountings to achieve optimum runner length and configuration. This "Short Ram" was tuned for an engine speed of 6000 to 6500 RPM. These factory headers, incorporated long, smooth bends, 3-inch collectors and a crossover connection.

Sometimes, the factories seem to equip an engine with an oversized carburetor. In fact, to some extent they do. Even though the installed carb may be a "bit" large for most street driving, it provides a decided edge in competition classes that require stock carburetion.

is required to keep the incoming air/gas mixture flowing after the piston has started moving upward (during the initial stage of the compression stroke) if the cylinder is to completely "filled," i.e., a 100% or greater v.e. is to be achieved. This force is called inertia, or a "ram" effect, and is a natural force accompanying any moving mass (in this case, the air/fuel charge) that tends to keep it moving once it has motion. Inertia increases with RPM, which is one of the reasons that internal combustion engines produce maximum horsepower in the upper RPM ranges.

Correlating air flow and volumetric efficiency theories to real life may seem problematical because theory and practice frequently do not coincide. When Detroit was producing high performance engines in the mid and late Sixties, the original equipment carburetors, in retrospect, appear to have been too large. Chevrolet used an 850cfm carb on the 427 Corvette and a 780 on the 302 cubic inch Z-28 Camaro; Ford used a 715 on the 289 and two 540cfm units on the 406 engine. Plugging these engine sizes and carburetor cfm values into our highly revered equation, the answer "does not compute." It would appear that the factories were expecting 8500+ RPM out of the smaller engines and 7500+ from the larger ones. Most stock engines would encounter valve float or suffer from internal damage before reaching these RPM levels, and nobody knows this better than the people who produced those engines. But 427 Corvettes, Z-28 Camaros, 289 Mustangs and the like were frequently entered in sanctioned competition where the original carburetor had to be retained. However, utilization of a carb that was too large for street use also served to insure that race engines were not undercarbureted. By the same token, engine options that were more carefully designed for the street were quite inappropriate for competition use, as they were invariably fitted with much smaller carburetors. Both the 390 horsepower 427, and 350 horsepower 327 Corvettes rolled off the assembly line with a 585cfm carburetor and the 428 Mustang used either a 675 or a 735cfm unit. (As further evidence of the air flow formula's accuracy, the larger carburetor was used only in 1968 and 1969 when the 428 Cobra Jets were raced extensively in Super Stock drag competition.)

AIR VALVE SECONDARIES

So, in the real world, it would appear that optimum carburetor size can in fact be calculated accurately. And as long as realistic RPM and volumetric efficiency figures are used, excessive over- or

Carburetor tuning is made somewhat easier by the good selection of jets, metering rods, inlet needle-and-seat assemblies and accelerator pump squirters from Carter. Rebuild kits are also available for "overworked" carbs.

under-carbureting should not be a problem. The only other basic consideration is secondary throttle activation method on four-barrel carburetors. Air valves (which are controlled by manifold vacuum) offer potentially smoother operation, and in theory "size" the carburetor to the needs of the engine. If the air valve spring or counterweight is properly tuned, maximum air velocity through the carburetor will be maintained at all operational levels. This is theoretically possible because secondary throttle opening responds to engine demands. Therefore, a 750cfm carburetor may never flow more than 675cfm, if that's all the engine requires.

Mechanical secondaries offer the further advantage of allowing the engine tuner to control when the secondary throttle bores reach maximum flow capacity. This is especially important in oval track and road racing where performance "coming out of the corners" is critical. Positive secondary throttle control is also advantageous in drag racing, particularly with automatic transmissions; the "tuneability" of the air valve has endeared the Carter AVS and Thermo-Quad designs to more than a few drag racers.

Because of the "tuneability" of the air valve, the AVS design is typically best suited for street, R.V. and recreational boat use, but mechanically operated secondaries are favored in virtually all competition applications. A notable exception to this generality is off-road racing; due to the inconsistency of the terrain, and lack of gas pedal control that arises from the driver ricocheting all over the passenger compartment, smooth application of power is of prime importance. If the secondary throttles are fully opened when the gas pedal is rudely smacked against the floorboard, excessive wheelspin can result, causing the rear tires to dig a hole rather than pushing the vehicle forward. Air valve secondaries alleviate this problem by opening more gradually and produce power more smoothly. Modifications of air valve opening rate are discussed in later chapters.

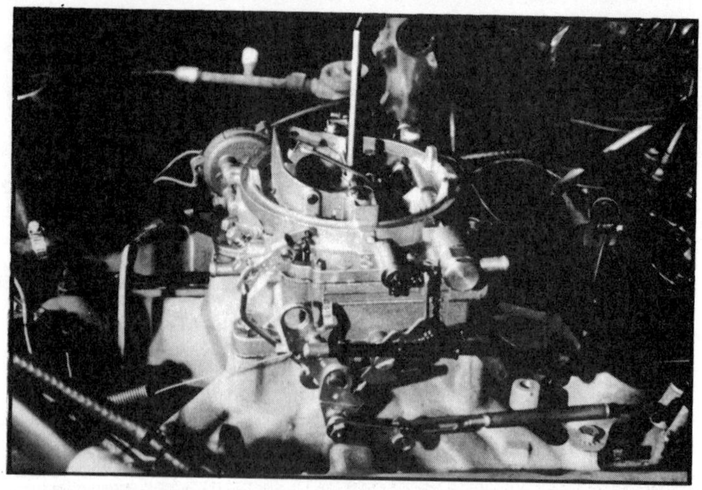

An intake manifold change will frequently alter the opening rate of vacuum-actuated secondaries—necessitating an adjustment. This is especially true when switching from a dual-plane to a single-plane type or when altering/restricting the heat-riser system.

CARB—MANIFOLD—CAM: A MEANINGFUL RELATIONSHIP

Carter offers an assortment of performance carburetors for most applications. Choosing a proper part number is therefore relatively easy, especially in light of the selection information and recommendation charts presented in this book. But there are occasions when "the right" carb simply doesn't afford the performance levels expected.

Irrespective of horsepower figures produced on a dyno, or through engineering theory or mathematical formulas, the bottom line of the performance ledger concerns a carburetor's ability to function in a "real-world" environment. And the overall functioning of this environment is determined in part by the intake manifold and camshaft. In many cases, a carburetor is expected to compensate for inadequacies in the intake system over which it has absolutely no control. A carburetor of different air flow capacity may alleviate some undesirable operational characteristics, but such solutions are often relatively poor compromises.

Not walking on a broken leg may assist the healing process, but the leg still must be set in a cast if it is to mend properly. Similarly, a carburetor cannot entirely compensate for the poor low-speed characteristics that are typical of long duration cams and/or large plenum single-plane manifolds. There have even been cases where over-enthusiastic folks have attempted to run 2x4 ram-type manifolds on street vehicles. But without extensive tuning of the carburetor/manifold/camshaft combination, results are usually disappointing. If you're convinced that your life depends on running a ram manifold on a street engine, let me not try to confuse you with the facts. Instead—some inexpensive advice. Make matters "least worse" and keep cam duration relatively short (220 degrees or less at .050-inch lift) and use two 400cfm carburetors on a mini-plenum ram manifold.

Like most cam grinders, Crane offers a wide selection of cams for virtually every popular engine. "Too much cam," like "too much carburetor," will make tuning more difficult and degrade driveability and gas mileage. Crane advises matching cam duration to engine speed at 55-60 miles per hour for optimum street performance.

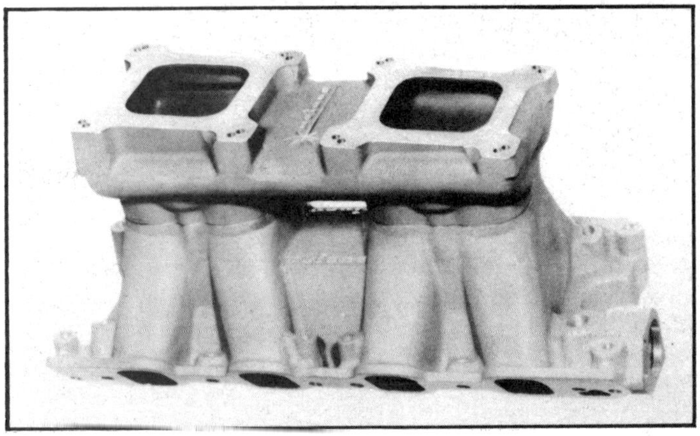

A dual four-barrel ram-type manifold is *NOT* the "trick" setup for the street. It can be made to work, but a single carburetor on a conventional manifold will almost always provide better overall performance.

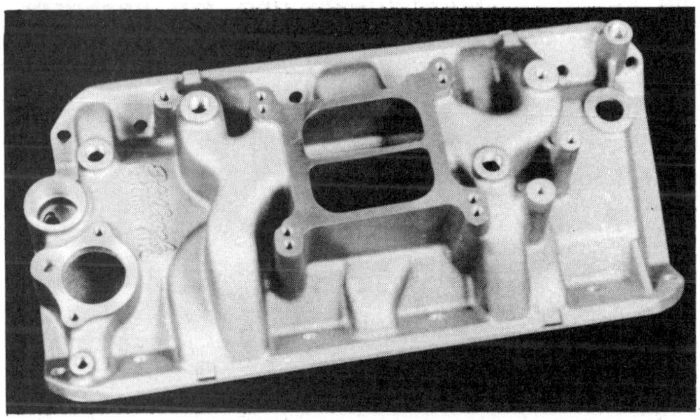

Stock and slightly modified engines will respond best to a dual-plane, high-rise-type manifold such as this Edelbrock Performer. Driveability is excellent, as is power up through 5500 RPM.

To better understand the problem of carburetor/manifold/cam mismatches, consider the case of a poor misguided street performance enthusiast. Seeking to emulate the heroics of his favorite racing idol, he purchases a long-duration camshaft, competition model single plane manifold, competition type carburetor and 1-7/8-inch diameter headers for his 350 cubic inch Chevrolet engine. Then, totally mystified by the fact that his car affords neither the fuel economy nor the performance he expected, he

attempts to improve his unfortunate situation. The first corrective measure is usually to install a smaller carburetor, in an effort to improve low-speed and mid-range response by increasing manifold vacuum at WOT. But the combination of a large-runner intake manifold, large-diameter headers and long-duration camshaft will consistently relay inaccurate fuel requirement information (vacuum signal) to the carburetor. Reduced vacuum at idle created by excessive cam overlap and poor exhaust scavenging will delay activation of fuel flow through the main nozzles, thereby creating an off-idle stumble. This can be rectified somewhat by increasing accelerator pump discharge volume, or by enrichening the idle circuit, but fuel economy will suffer. A more satisfactory approach would be to install a shorter duration camshaft, and a dual-plane manifold or appropriate equivalent.

REAL LIFE AND OTHER CONSIDERATIONS

Whenever a carburetor is purchased, the buying decision should be made based on an honest appraisal of the equipment with which the carb must interact and the RPM level at which the engine will *realistically* operate. Assuming that the cam, manifold and headers in the example above had already been installed, a smaller carb with air-valve secondaries would have provided better (although not necessarily satisfactory) driveability. (Even on a race engine, excess air-flow capacity can reduce performance to below optimum levels.) But there is more to carburetor selection than mere size consideration. Certain economy/emissions carburetors are fitted with fuel-metering circuits that are *very* sensitive to changes in manifold conditions. When these carburetors are used in conjunction with a long-duration camshaft, the pulses (in the intake manifold) generated by excessive valve overlap can degrade fuel metering capabilities. Once again, a single-plane manifold with a large open plenum intensifies such problems.

Although most carburetors offer acceptable performance when installed on a single-plane manifold, if cam duration (measured at .050-inch lift at the cam) is in excess of 230 degrees, inherently low manifold vacuum at idle may require that a dual-plane or "Z" type manifold be used to afford reasonable low-speed driveability and throttle response. The special economy type manifolds such as the Edelbrock SP-2P and Holley "Z-Series" are specifically designed for stock or mildly modified engines. Therefore, these manifolds may complicate tuning procedures and compromise engine performance when used in conjunction with large carburetors, long duration camshafts or race type headers.

When economy and low-speed operation are the top priorities, a manifold such as the Holley "Z-Series" is a good choice. The design was created by Zora Arkus-Duntov, the father of the Chevrolet Corvette.

CHAPTER 4 CARBURETOR FUNCTION

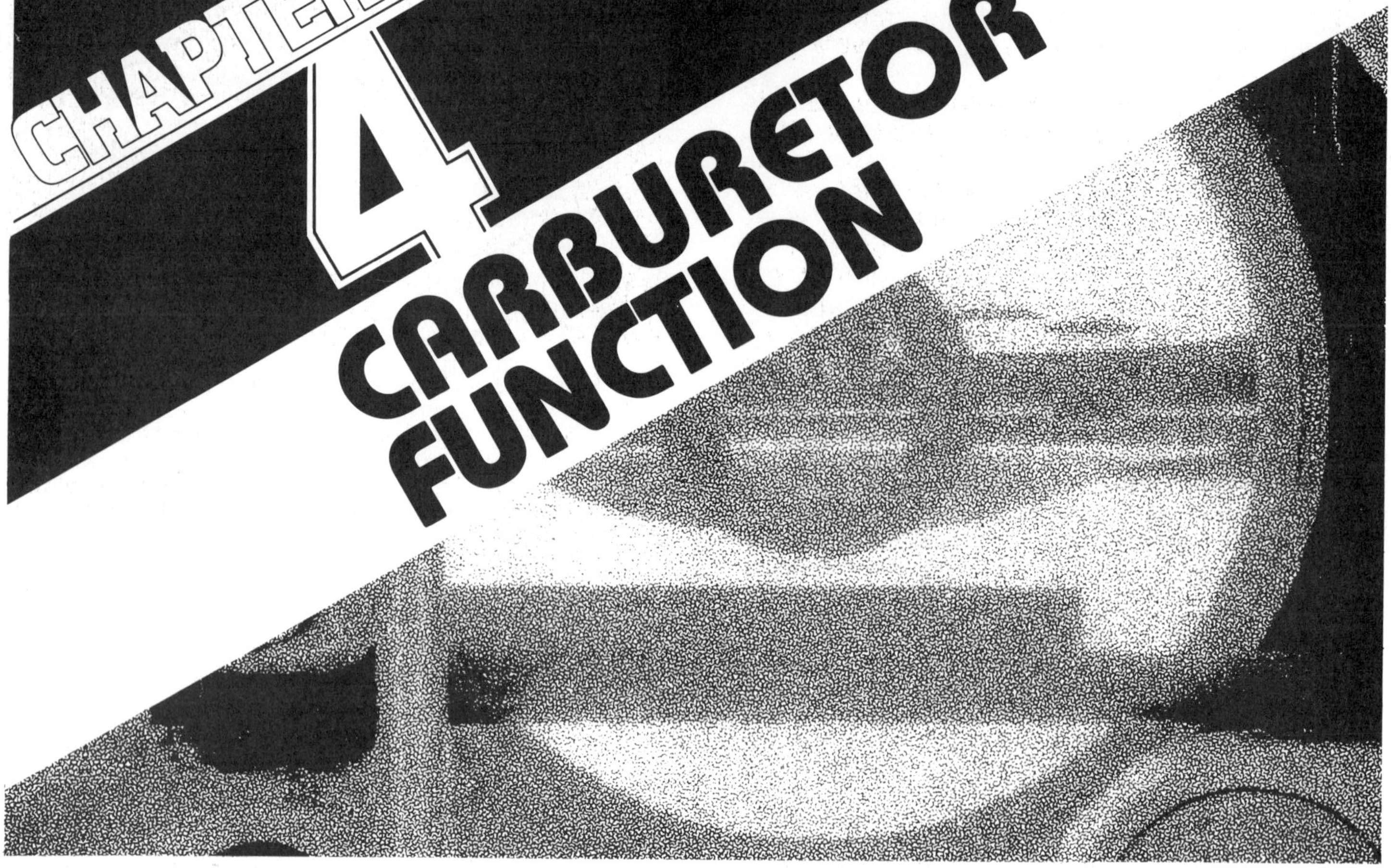

Compared to an AFB, this Carter model U-1, which was the company's first down-draft model, looks primitive at best. Modern carburetors are certainly more sophisticated than this 1931 model, but the basic venturi function is very similar.

CARTER COMPLEXITY: MYTH OR FACT?

In the days of the Model-T Ford, a mechanic didn't need a great deal of expertise to be considered a "carburetor expert." The unsophisticated single-barrel fuel mixers used in those days had few parts, were easy to disassemble and suffered primarily from problems induced by impurities contained in the gasoline which they crudely metered. Today, the situation is markedly different. Multi-barrel carburetors have achieved an undeserved reputation for being difficult to deal with, and as impossible to comprehend as a politician's logic. And even

To the uninitiated, this is just an ordinary Dodge Charger, but the educated eye will immediately detect the "HEMI" insignia on the side. The Chrysler Hemi did much to perpetuate the aura of mystery that surrounds the Carter AFB.

though information is more readily available than it was in the early days of the automobile, the current myriad of carburetor designs and models has generally served only to muddy the waters of understanding.

Part of the reason that carburetors in general, and Carters in particular, are reputed to be complicated, has to do with the fact that the AFB, having been the original equipment carburetor on such legendary engines as the 409 Chevrolet, Chrysler Street Hemi and Pontiac GTO, has always been shrouded by the Muscle Car Mystique. In truth, neither the AFB, AVS nor Thermo-Quad are difficult to comprehend.

Except for a few eccentric designs, which were originated by equally eccentric engineers, all modern automotive carburetors are very similar in concept. Moreover, they are identical in basic function; differences lie in the methods used to accomplish those functions. However, Carter carburetors, not being of a modular design, do not fully bare their most intimate fuel circuits, even when completely disassembled. This does tend to raise the confusion factor, but with a few cutaway drawings, there's clear sailing ahead.

BACK TO BASICS

The basic purposes of any carburetor, be it a one-, two- or four-barrel, are atomization of fuel, mixing of air and atomized fuel in the amounts required by engine load, and regulation of engine speed. These concepts can be understood most easily by exploring the theory of carburetion (the mixing of fuel with air) and then relating the theory to functional hardware as incorporated in various Carter models.

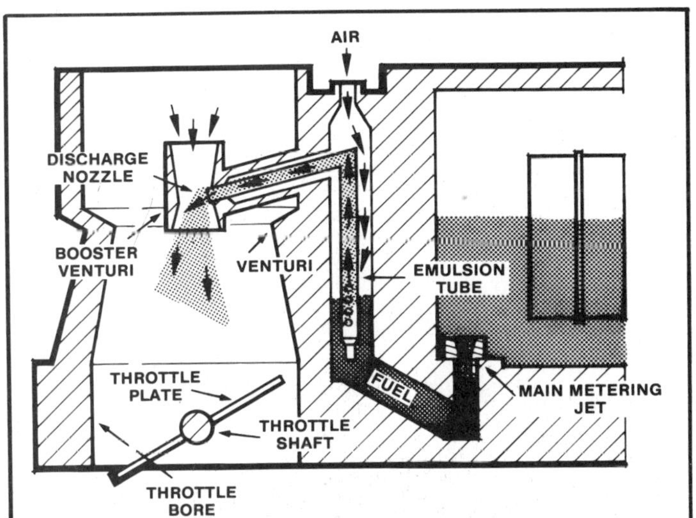

BASIC MAIN METERING SYSTEM
AIR AND FUEL PATHS

Air is introduced into the fuel at two distinct points in the typical carburetor. First, air metered through an air bleed is drawn into the fuel in the emulsification tube. Then additional air is mixed with the emulsified fuel in the booster and main venturis. When fuel reaches the intake valve it is almost completely vaporized.

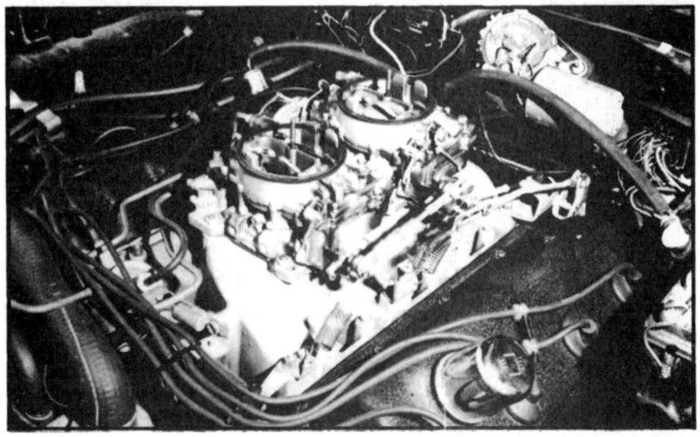

The front AFB on this factory Hemi installation has no provision for idle-mixture adjustment. Under this arrangement, the rear carb is considered a "primary"—with an idle circuit—while the front carb serves as a "secondary." Since the front barrels of the rear carburetor are centrally located, fuel distribution is acceptable.

In order for gasoline to burn within an internal combustion engine, it must be rendered into a near gaseous state by thoroughly mixing it with some amount of air. Air is necessary to provide oxygen, a requisite of combustion. Prior to the days of stringent emission requirements, 14.7-14.8 parts of air (by weight, i.e., pounds) to one part of gasoline was thought to be the optimum mixture ratio for internal combustion engines, but current technology has enabled engines to run as lean as 17:1 or 18:1. Under full throttle heavy-load operation, a richer mixture (in the vicinity of 12:1 or 13:1) is generally desired, for this richer air-fuel ratio will produce maximum horsepower.

Efficient combustion of gasoline is only possible when the fuel is mixed with 13 to 17 times as much air (by weight). The mixing, performed remarkably well by the relatively simple automotive carburetor, accurately and uniformly meters small quantities of fuel into a rapidly-moving airstream.

Suction, caused by the pistons moving down their bores during the intake cycle, creates a partial vacuum in the induction system. In turn, these low pressure pulses are communicated through the intake manifold to the discharge nozzle and cause fuel to flow into the air stream. At this point, the fuel, pouring into the fast moving air, is atomized, (converted to a spray of fine droplets). It has not necessarily been vaporized. It remains in small droplets until it reaches the low pressure (high

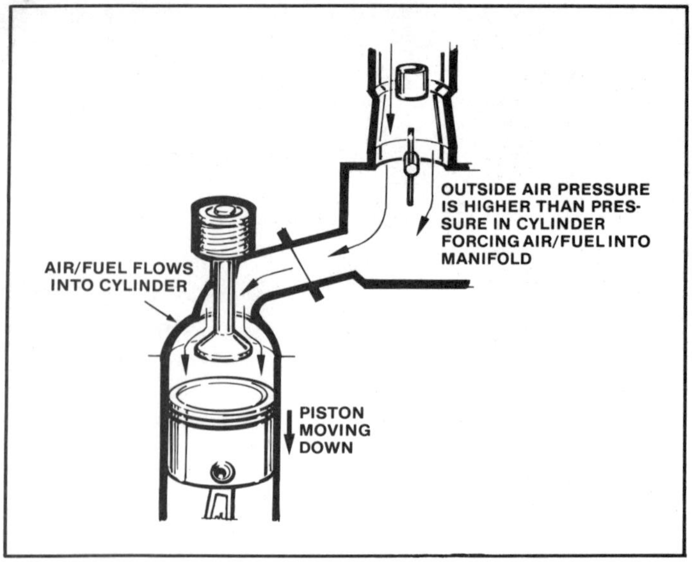

As the piston moves down the cylinder during the intake stroke, a lower pressure is created in the combustion chamber, intake port, and manifold runners. Air from the outside rushes in the carburetor to equalize this pressure differential, drawing the air/fuel mix into the cylinder.

vacuum) area beneath the throttle plate that enables it to "boil" and become a gas. Strong manifold vacuum is a necessity, not only for good atomization, but also to provide a proper (low pressure) environment for vaporization, essential for both peak efficiency and maximun performance.

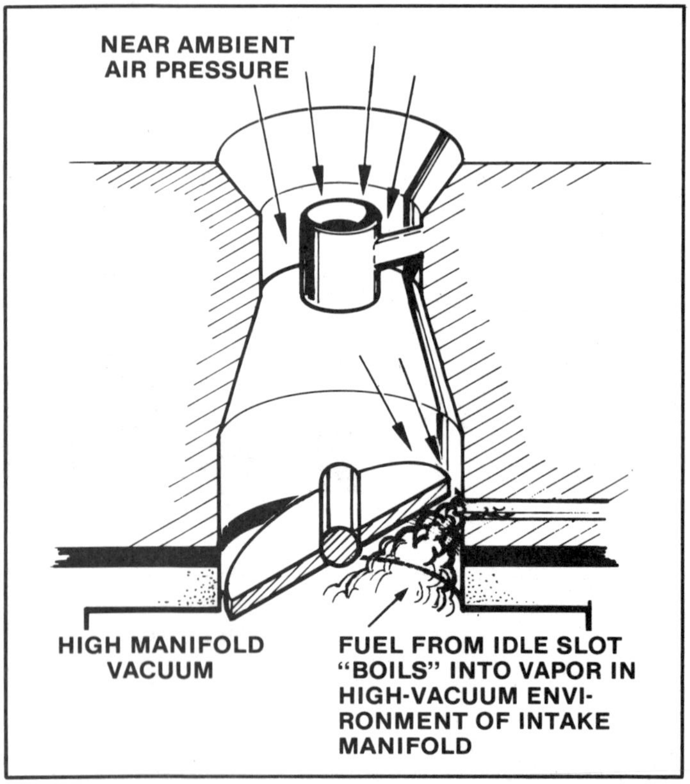

When fuel droplets, in this case from the idle slot, enter the low pressure (high vacuum) area below the throttle plate, they "boil" into a near complete vapor and thoroughly mix with the surrounding air.

PRESSURE OR VACUUM?

This concept may be more readily understood if compared to an automobile cooling system. Under normal circumstances, water boils at 212° Fahrenheit, (100°C) but automotive coolant temperatures frequently exceed that level without the contents of the radiator turning to steam. Were it not for the pressure cap, this would not be the case. (For the purpose of this discussion, let us assume that the entire cooling system is filled with water rather than a water/anti-freeze mixture.) Pressure keeps the water in a liquid state at temperatures where it would normally turn to steam.

But if the pressure is relieved, even a small amount, when water temperature exceeds 212°F, steam will hiss out the overflow tube. Similarly, when gasoline flows from above the throttle plate (where pressure is close to atmospheric, i.e., 14.7psi) to the lower pressure environment of the intake manifold, the gasoline "boils," changing from a liquid to a vapor (gas). Since manifold vacuum is typically measured in inches of mercury, (in/Hg) it must be converted to psi in order to relate the cooling system analogy to the pressures existing above and below the throttle plate. Assuming a manifold vacuum of 18in/Hg, manifold pressure may be computed as follows:

Standard atmos. psi @ sea level = 14.7
PSI=in/Hg x .491(conversion factor)
Therefore:
PSI=18 x .491=8.83psi
Or:
18in/Hg=8.83psi

Since manifold vacuum is a measurement of the psi difference below atmospheric pressure, 8.83psi must be subtracted from normal atmospheric pressure (of 14.7psi) in order to determine the residual pressure in the intake manifold.

Atmos. pressure - manifold vacuum=manifold pressure
Or: 14.7psi - 8.83psi=5.87psi

This computation demonstrates that a partial rather than a complete vacuum exists within the manifold. There is some pressure present, but it is substantially below atmospheric. Therefore, when gas droplets travel from above to below the throttle plates, they are in fact moving from a high to a low pressure environment (14.7 to 5.87psi). Obviously as manifold vacuum decreases (this occurs when the throttle is opened) the pressure differential is reduced and vaporization is not as complete. But the pressure differential is never totally eliminated. If it were, air and fuel would not move through the carburetor.

MR. VENTURI AND HIS PRESSURE DROPPER

If an engine were designed to operate under a

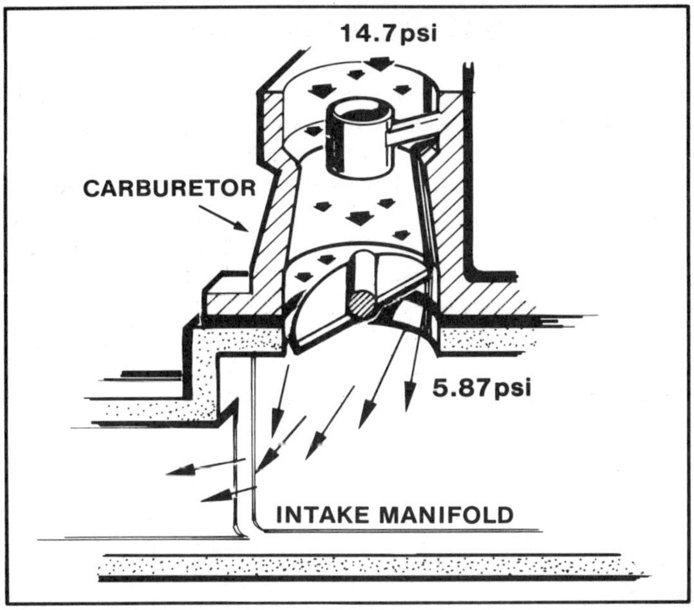

The area below the throttle plates is not a complete vacuum, but rather contains a lower-than-atmospheric pressure. As our text example illustrates, with a manifold pressure differential of 8.83psi (below atmospheric pressure of 14.7psi) the resultant manifold pressure is 5.87psi. This pressure differential is reduced, but never eliminated, as the throttle is opened.

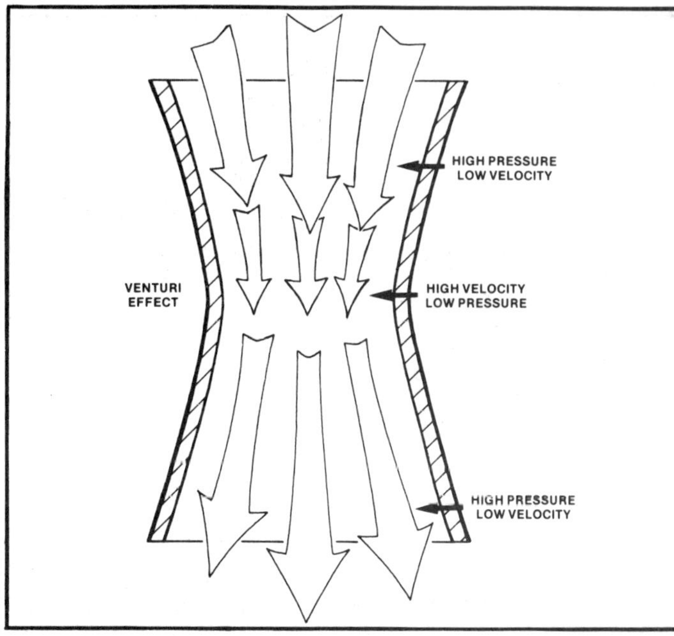

Carburetor function is based on the Venturi Effect. As air moves through a venturi, velocity increases and pressure decreases. The pressure decrease "draws" fuel into the airstream. The volume of fuel is controlled by the restriction of a jet and metering rod combination. This simple design is the basis behind the function of all automotive carburetors.

constant load, at a constant RPM (this is not uncommon in industrial engines), the throttle opening would be permanently set and vaporization conditions would be ideal since manifold vacuum and air velocity (through the carburetor) would never vary. The carburetor for such an engine would need only a single fuel metering circuit, and it would be similar to the one that is commonly referred to as the main system. Mixture (air/fuel ratio) would be controlled by a nonchangeable jet, and fuel would be discharged only through a nozzle located in the low pressure portion of the venturi.

Over 150 years ago, G.B. Venturi, an Italian scientist who had no interest whatsoever in automobiles, discovered that when air moves through an hourglass shaped tube, (diameter changes from large to small and back to large) the velocity is highest and pressure lowest in the area of smallest diameter. This principle, named for its discoverer, forms the basis for virtually all carburetor design. The low pressure condition found in the venturi, draws fuel out of a reservoir (float bowl) through the discharge nozzle, which is located in the narrow (low pressure) region of the venturi.

With low pressure in the venturi acting in the same manner as a suction pump, there is no reason for fuel to exit the discharge nozzle in anything other than a liquid state. Just stick a straw into a glass of vintage wine (rather gauche, but it serves the purpose) and suck on the other end. The principle is identical. Suction is merely a pressure reduction that serves to initiate flow; it provides no means of converting a liquid (be it wine, gasoline or anything in between) to a spray mist. Atomization results when a stream of liquid is introduced into a column of fast moving air, but without an additional step, many of the gasoline droplets would be too large to be vaporized and would remain liquid (and for the most part, unburned) in the combustion chamber. The additional operation needed in the liquid-to-mist

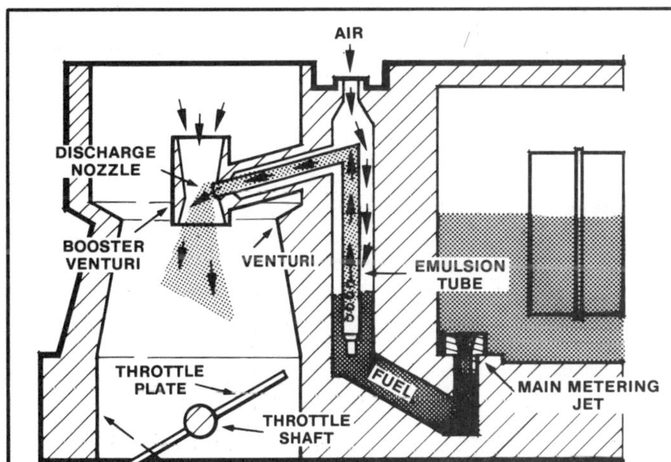

BASIC MAIN METERING SYSTEM
AIR AND FUEL PATHS

If raw fuel were discharged from the main nozzle, a large portion would enter the combustion chamber as a liquid and would not burn. But when fuel is "pre-atomized" in the emulsification tube it is nearly vaporized by the time it reaches the intake valve, resulting in more complete combustion, better performance and gas mileage, and fewer emissions.

breakdown process is accomplished between the fuel bowl and discharge nozzle, and is known as emulsification or pre-atomization.

WHEN AIR MEETS GAS

According to dictionary definition, an emulsion is "a mixture of mutually insoluble liquids in which one is dispersed in droplets throughout the other." If one were to take quantities of vinegar and vegetable oil, pour them in a bottle and shake vigorously, the end product would be an emulsion (one that we commonly call salad dressing). If the emulsion is allowed to stand, the two liquids will soon separate and the lighter substance will rest entirely on top of the heavier one.

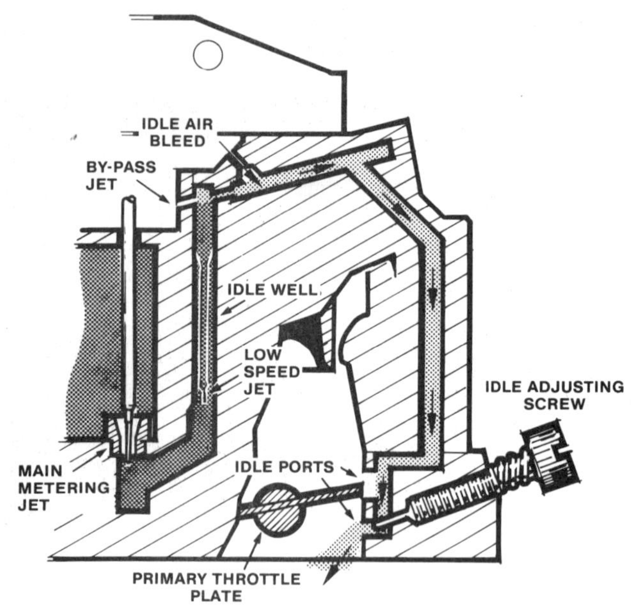

IDLE AIR/FUEL EMULSION

Idle air/fuel is emulsified by a separate circuit. Air is usually added to the idle fuel through an idle air bleed. The external idle mix adjusting screws control the overall flow of the emulisified mixture, while a restriction in the idle well (here labeled the "low speed jet") limits maximum "fuel-only" flow.

In the world of carburetion, an emulsion refers specifically to droplets of fuel interspersed with a volume of air. The device that is most instrumental in fuel emulsification is known as the air bleed. The air bleed is an air metering jet in the end of a passageway connecting with the fuel delivery line to the discharge nozzle. The same suction (low pressure) force that causes fuel to flow to the discharge nozzle, now draws air into the liquid fuel stream (through the air bleed). The effect is similar to the experience of drawing liquid through a straw that has a small hole in its length. The liquid moves up through the straw, but it reaches your mouth in a bubbly froth. So in a sense, when you drink the wine used in the previously mentioned suction experiment with a straw, that has such a hole, you're drinking emulsified Chenin Blanc, and it is no different than the air/fuel mixture that exits at the main discharge nozzle (with the exception of taste, of course).

LET IT BLEED

However, the function of an air bleed is not solely to emulsify fuel. It also exerts control over fuel flow by "bleeding off" some of the suction force or fuel metering "signal" that is developed at the discharge nozzle. By varying the size of the bleed, the amount of suction (vacuum) required to start fuel flow can be altered. As bleed size is enlarged, the vacuum required to initiate fuel flow is increased. Conversely a reduction in bleed size reduces vacuum signal requirements and fuel flow is more easily initiated.

AIR/FUEL MIXING (EMULSIFICATION)

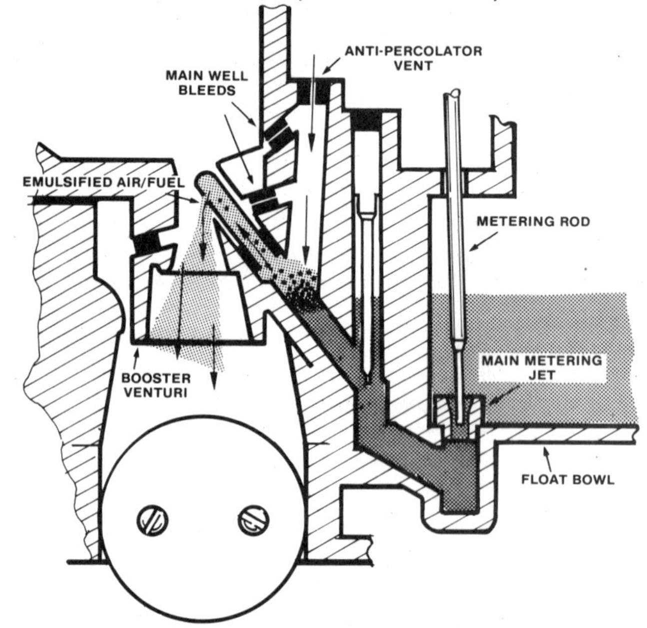

The exact technique for emulsification will vary from one carburetor model to the next. However, the principle of drawing air into the fuel flow by the vacuum created in the main and booster venturis does not change.

Let's look at an extreme example of this concept, as an aid in understanding this important functional property of carburetors. If the bleed could be infinitely enlarged, the point would be reached where a suction pulse was never transmitted to the fuel in the float bowl. In essence, the entire signal would be bled off and only air, entering through the air bleed, would flow through the circuit. Total elimination of the bleed, on the other hand, would allow transmission of the full signal to the fuel reservoir. While this might seem like an excellent means by which to instigate fuel discharge at low manifold vacuum, lack of an air bleed would lead to percolation problems.

In this view of an AFB, several small holes in the booster venturi castings are visible (arrows). These are air bleeds, and as explained in the text, they play an important role in determining carburetor air/fuel ratio characteristics.

Air bleed diameters are determined by applying solid engineering and sophisticated test equipment, like this air flow bench. Air bleed diameters are extremely critical since they affect not only air/fuel ratio, but circuit timing.

The booster venturi in the secondary of this AFB creates a high pressure drop around the discharge tube, initiates fuel flow earlier and improves low-speed performance. Carter carburetors are often equipped with one or two (double) booster venturis.

After an engine has run for some time, the carburetor body becomes extremely warm, and on occasion, down right hot. When the engine is switched off, carburetor heat can temporarily increase since fuel flow stops (the exiting fuel carries off heat with it, as does the engine fan). The fuel remaining in the float chamber begins to absorb heat from the surrounding carburetor body and in some cases, the fuel will actually boil or percolate, much like water in a coffee pot. Without an air bleed or anti-percolator vent, hot fuel remaining in the main wells could be forced out of the discharge nozzle by the vapor pressures that result from the percolating action. The presence of a vent allows pressure in the main well to escape thereby preventing fuel from being forced out of the discharge nozzle.

The air bleeds contained in production carburetors are precisely measured restrictions, sized such that fuel flow from the discharge nozzle into the air stream is started at specific vacuum levels. (It should be noted that this discussion pertains to venturi, not manifold vacuum. Venturi vacuum increases with air flow and will therefore be greater at wide open throttle, than at idle or part throttle. This, of course, is in direct contrast to manifold vacuum which decreases as the throttle is pushed toward a wide-open position.) Without the sophisticated equipment required to perform fuel flow analysis, it is impossible to accurately assess the results of bleed size alterations. This is why most carburetor engineers advise against such modifications.

It should also be obvious that bleed size affects air/fuel ratios. A large bleed will admit more air than a small one, and consequently a leaner mixture will result.

BOOSTER VENTURIS—FOR A STRONGER SIGNAL

At this juncture, it should be clear to anyone who has looked down the throat of a Carter four-barrel that something more sophisticated than a simple "dump tube," mounted in a low pressure area, is required for satisfactory performance on a modern automotive engine. With a mere tube protruding into the venturi, serving as the discharge nozzle, inordinately high air flow is required to develop a sufficient metering signal to initiate fuel flow. The restriction created by a standard venturi will simply not provide a large enough pressure drop (in the immediate vicinity of the discharge nozzle) to cause fuel flow at low air velocities. In essence, the pressure drop (and resultant fuel metering signal) needs to be boosted, and the device utilized to perform this function is appropriately called a boost or booster venturi, and it acts as a venturi within a venturi. There really is a method to all this carburetion madness.

The use of a booster venturi makes the carburetor "look" smaller without greatly affecting air flow. In addition, a booster offers a number of advantages, although they pertain primarily to production considerations. Rather than requiring a different main body casting (with a venturi of a particular diameter) to "fine tune" a carburetor for flow capacity and signal strength, one casting can be employed and the shape and size of the booster may be altered. The cost of changing a relatively simple booster is far

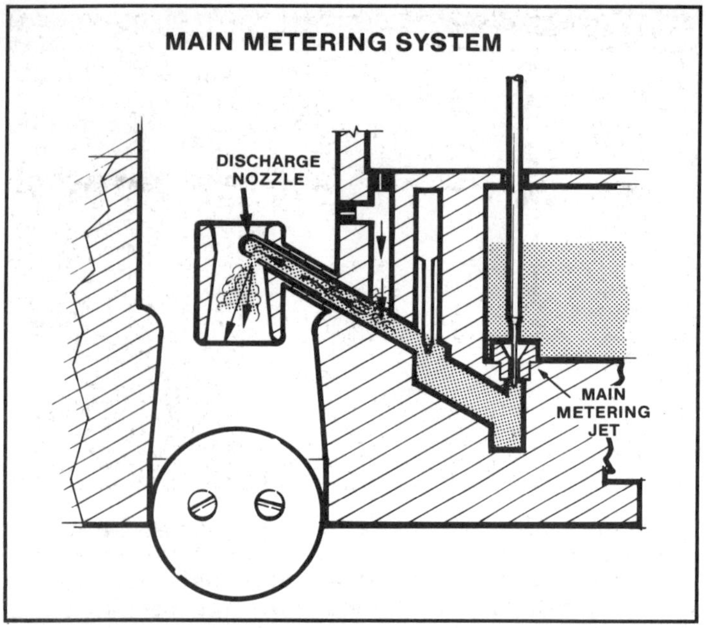

The booster venturi creates additional "signal" and initiates fuel flow at lower vacuum/air-flow conditions. The main restriction in the primary high-speed circuit (the main metering jet/metering rod combination) controls the overall air/fuel ratio. Together they form a precision "feedback" circuit that tailors fuel flow to engine requirements.

less than the expenditure necessary to alter an intricate main body casting. However, this "tuning" capability can only be applied over a relatively narrow range of air-flow capacities.

Another advantage of the booster concept is that vacuum can be intensified in the area immediately adjacent to the discharge nozzle. By varying booster shape, it is possible to increase signal strength while minimizing loss of air-flow capacity. In some carburetors, a double booster is utilized as a means of further intensifying the fuel metering signal.

Some carburetors, such as the Carter Thermo-Quad, are fitted with dual booster venturis—illustrated by this underside view of the primaries (arrows 1)—that provide a greatly intensified fuel metering signal. It is interesting to note that the Thermo-Quad also has no boosters—or venturis—at all on the secondary side (arrows 2). Location of the secondary air valve above the main system discharge nozzles (arrow 3) is necessary to create the pressure drop required to initiate fuel flow.

However, any booster design is limited in its ability to distribute fuel evenly around the throttle bore. When poor (on specific original equipment applications where poor) manifold designs compound this fuel distribution problem, Carter engineers frequently add a tab or mill off a portion of the booster's trailing edge. These modifications "shape" the low pressure area such that exiting fuel is pulled into a more desirable distribution pattern.

At this point, it should be noted that the Carter AVS and Thermo-Quad carburetors do not have a venturi of any type in the secondary barrels. Fuel is discharged through a simple dump tube that protrudes into the air stream. In spite of appearances, this is not contradictory to our previous discussion on venturi function. The concept works because it is employed only in the secondary barrels, where the throttle plates are operated mechanically and the area surrounding the discharge tubes is subjected to *manifold* vacuum, because it is "sealed" from the atmosphere by an air valve held closed by an adjustable spring. Therefore a fuel delivery signal is presented to the discharge nozzles *before* air begins to flow in any appreciable volume. In this

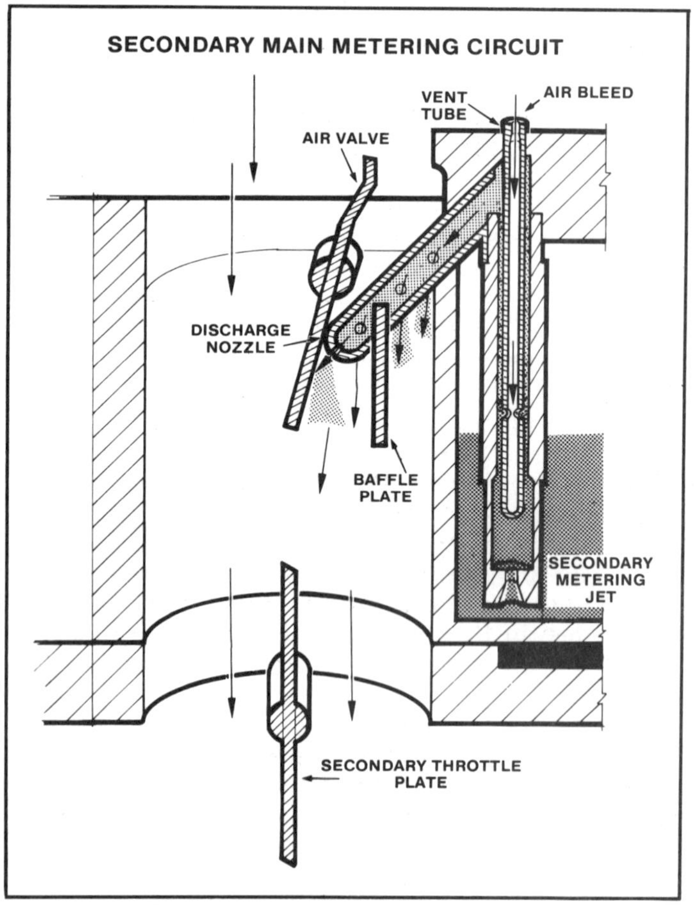

The air-valve secondary uses vacuum created under the air valve to initiate air/fuel flow. The design of the air valve, discharge nozzle, and baffle plate acts similarly to the booster venturi in the primary bores; providing a smooth transition from closed to full-throttle performance.

instance, manifold vacuum, not venturi vacuum, causes fuel to flow out of the nozzle. Were it not for the air valve being located above the discharge nozzle, fuel would not flow at low speed and the engine would suffer from an extreme lean condition as the throttle is opened.

APPEARANCE VS REALITY

With the changing priorities of recent years, and the EPA's grand and glorious entrance into the automobile business, fuel distribution became especially important. It is for this reason that some carburetors, when used both as original equipment and for aftermarket replacement, differ, even though they share the same model designation. New car emissions standards may dictate internal changes not required on performance-type replacement carburetors.

Since each individual booster design creates a vacuum signal of different intensity, it can be seen that fuel metering requirements and air bleed dimensions can vary significantly between carburetors, even among those with identical throttle bore and venturi diameters. And by dealing with these variables, carburetor engineers can create an almost infinite variety of fuel delivery curves.

JETS—THE KEYS TO MIXTURE CONTROL

Without the proper equipment, altering the air bleeds or booster venturis is tantamount to playing with fire (which can be quite dangerous around carburetors!). These items are simply not designed to be modified, and misdirected enthusiasm can ruin an otherwise functional carburetor. However, the metering jets can be removed and replaced by anyone who understands the operation of a screwdriver. The jet is therefore the means provided for altering fuel flow to suit a specific requirement.

A carburetor jet is nothing more than a metering device that controls the flow of fuel from the fuel bowl into the main circuit. After passing through the jet orifice, fuel travels into the main well, where it is emulsified, (mixed with air from the air bleeds) before moving on to a connecting channel which leads to the discharge nozzle. In most Carter models, the jets are located in the float chamber floor. Fuel must travel through a channel located beneath the jet before reaching the main well.

On Carter four-barrels, the air bleeds typically function in combination with an anti-percolator vent, however, specific vent and bleed placement varies with individual carburetor models. For information relating to placement on specific models, refer to the appropriate chapter in this book.

Irrespective of bleed placement, it is the vacuum (pressure drop) created by the booster and main venturi—or in the case of the secondary air valves, exposure of the nozzles to manifold vacuum—that simultaneously causes air to be drawn through the air bleeds and fuel to be drawn through the main jets.

In all Carter four-barrel carbs, (AFB shown here) the primary jets reside in the floor of the float chamber. They can be easily removed and installed, but to prevent jet damage and metal shavings from entering the main well only use a wide-bladed screwdriver that snugly fits the screw slot.

And it is the purpose of the main metering circuit to mix fuel and air in the proper proportion.

POWER ENRICHMENT CIRCUIT

So long as the fuel metering signal remains constant or increases in intensity (for a corresponding increase in vacuum) fuel flow will be sufficient to maintain the air/fuel ratio as determined by the metering jet/air bleed/venturi relationship. But what happens when signal strength drops, as often happens when the throttle plates are opened quickly? Obviously, a leaner air/fuel mixture will result; as vacuum decreases, the amount of fuel drawn through a fixed orifice (the jet) also decreases.

Except for the Thermo-Quad, Carter secondary jets are located in the float bowl floor and are similar in appearance to those used in the primary circuit. On AFB, AVS and WCFB carbs with two-step metering rods, the primary and secondary jets are actually interchangeable. However, AFB and AVS carbs fitted with three-step rods require special "high type" primary jets.

Consequently the mixture becomes lean—at precisely the time when acceleration demands additional richness.

Engine vacuum also drops during periods of extended high load—for example, when climbing a hill or towing a trailer—so a carburetor must be designed to meter additional fuel for all heavy load conditions, irrespective of how long that load may be applied. Depending on your point of view, additional fuel for high load conditions is added by either a power enrichment or economizer circuit.

ECONOMIZER OR POWERIZER?

In days of old, (and occasionally in days of new) carburetors were said to have an "economizer circuit." This nomenclature is somewhat backwards since it is an apt description of the circuit only when it is not enabled. The circuit is actually designed to enrichen the fuel when needed, not to further lean the mixture, as might be suggested by the name. "Economizer" is undoubtedly the product of a brilliant marketing mind who was attempting to justify the added expense of another fuel circuit to a parsimonious, all-conquering accounting department. (Being given more to extravagance than frugality, I'll take the positive approach and empahsize power.)

As used in most Carter carburetors, (and in some other makes and models) the power enrichment circuit consists of a spring-loaded piston to which is attached a tapered metering rod (also called step-up rod). The metering rod is designed to protrude down into the main-jet orifice, thereby limiting fuel flow. During periods when engine vacuum is high, as during highway cruise, the piston is pulled to its full-downward position against the return spring. This movement allows the largest diameter section of the metering rod to enter the main jet thereby limiting fuel flow (as opposed to an unrestricted jet) and establishing the desired air/fuel ratio for light load conditions. Conversely, as vacuum drops below the point where spring pressure is overcome, the piston moves upward, pulling the large diameter part of the rod out of the jet allowing additional fuel flow through the main jet.

To simplify matters somewhat, operation of the power enrichment circuit may be likened to that of the automatic door opening systems found at most supermarkets. All you need do is approach the entrance, and like Moses parting the Red Sea, you cause the doors to fling themselves out of your path. Control of door opening and closure is handled by a switch that is activated by your body weight. The presence of weight trips the switch, causing the doors to open; removal of weight deactivates the switch allowing the doors to close. In effect, the switch senses environmental conditions and causes a response to change.

A piston that controls metering rod position can also be thought of as a sensing switch that controls a response. When manifold vacuum is high, there is no need for supplemental fuel flow, so the "doors" stay closed. But when vacuum drops, the switch is activated, thereby raising the metering rod; and with the "doors" open, additional fuel can flow into the main well. As vacuum again increases, the switch resets and allows the "doors" to return to their normal closed position, once again limiting fuel flow.

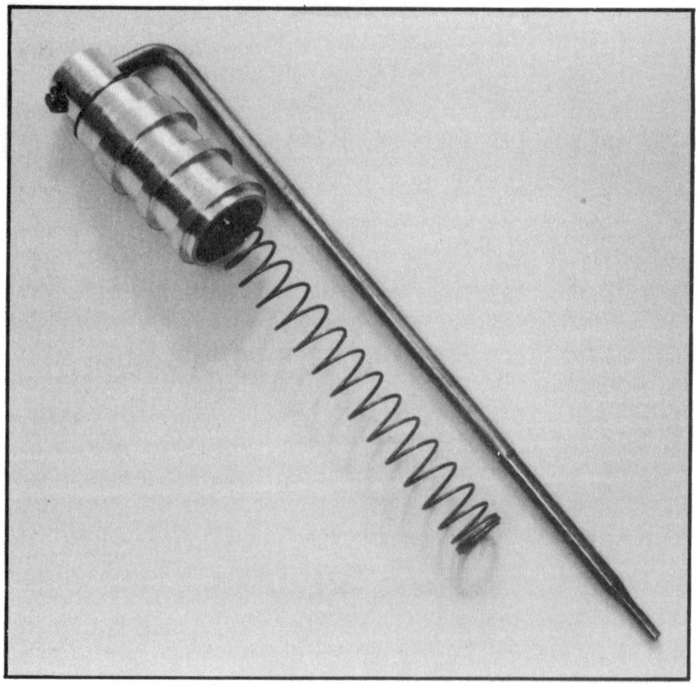

Carter uses stepped metering rods that move up and down in the jet orifices, introducing a variable restriction to fuel flow. This variability is designed to enrich the air/fuel ratio when performance is required and reduce fuel flow under light throttle cruise. The rods, which may have two or three steps of different diameters, are attached to spring-loaded pistons which react to manifold vacuum; matching fuel flow to engine requirements.

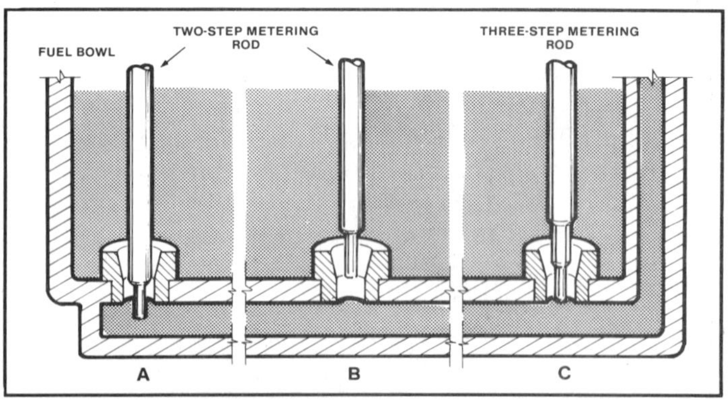

The metering rod position modifies the air/fuel ratio to suit engine requirements. At low speed, high vacuum, (A) the metering rod is fully inserted in the jet and leans the mixture to provide good mileage. During heavy-throttle, high-speed operation, (B) the metering rod is retracted from the jet and enrichens the mixture providing maximum performance. Three-step metering rods (C) are used in some carburetors to smooth the air/fuel ratio transition from part to full throttle.

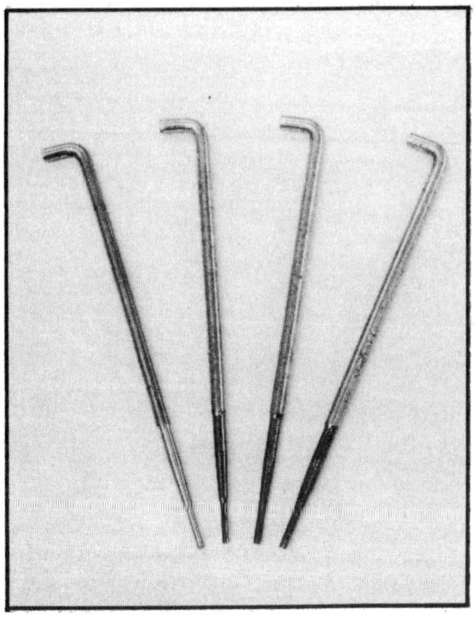

At first glance, all metering rods may appear similar, but Carter has used both two and three step models. The latter has an intermediate step for partial enrichment during light-load conditions. Metering rod diameter must be coordinated with jet diameter to insure proper air/fuel ratio for all engine requirements.

SPARE THE RODS

Another method of providing a richer mixture for high-load/full-power operation uses a so-called power valve, rather than metering rods. And while power valves are used in some Autolite, Ford, Holley, Rochester, Stromberg and Weber models, Carter engineers universally prefer the metering rod approach. But irrespective of the mechanics of operation, the power enrichment circuit is charged with adjusting the air/fuel ratio relative to manifold vacuum. With a power valve serving as an "extra-fuel switch," the amount of fuel supplied by the power enrichment circuit is controlled by a precisely sized orifice (the door size) located in or near the valve. Whereas, in circuits containing metering rods, the diameter of the main jet and the metering rod determines high-load fuel-flow capacity.

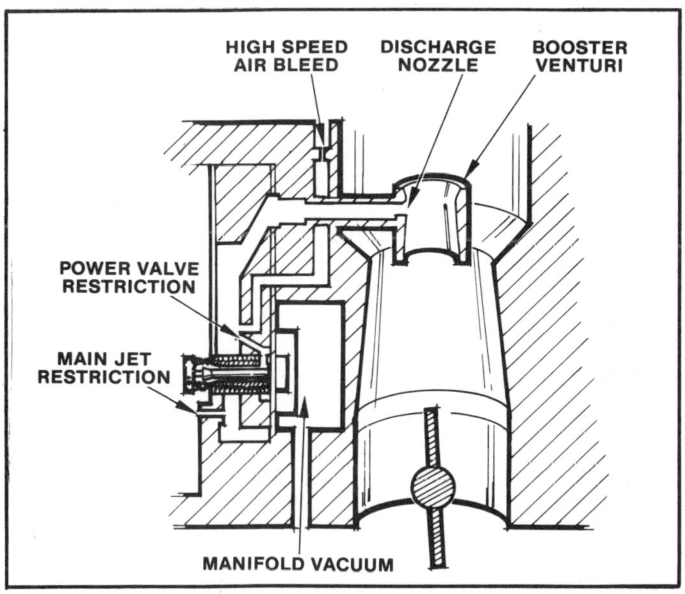

The power enrichment (power valve) system admits additional fuel to the main well when manifold vacuum declines—under high speed or heavy load. This added fuel is not metered by the main jet but by another built-in restriction called the PVCR (Power Valve Channel Restriction).

Implementation of a power valve actuated enrichment circuit is quite straightforward. When vacuum drops to a sufficient level, spring pressure working against a diaphragm opens the power valve and allows additional fuel through orifices, generally known as PVCR's (power valve channel restrictions). These restrictions meter the flow of fuel into the main well and consequently control the amount of fuel enrichment admitted through the main circuit.

A power valve circuit can suffer from a rupture of the rubber diaphragm (not uncommon), and leaks between adjacent passages. However, a metering rod design is virturally impervious to the effects of time, wear, and engine backfire. Failures in metering rod enrichment circuits are exceptionally rare, and are usually the result of the piston becoming seized in its bore or blockage of the passage that allows manifold vacuum to control piston position, often from a long-overdue rebuild. In both cases malfunctions are normally corrected by removing excessive carbon and varnish build-up.

MULTI-STAGE ENRICHMENT FOR OPTIMUM ECONOMY

In the early 1970's, when the popularity of recreational vehicles surged, there arose a need for multi-stage power enrichment. Carter was able to accommodate this requirement with metering rods that incorporated three steps in their diameter, instead of the usual two. Since rod movement is directly related to manifold vacuum, the addition of an intermediate step and the resultant partial enrichment improved fuel economy and driveability under moderate load conditions. Partial enrichment, at vacuum readings of 8.5-12.5in/Hg, allows jetting to be somewhat leaner than would be possible with a conventional two-step rod, as additional fuel is available for operation under light-to-medium load. This "improvement" is especially welcome in campers, motorhomes and other heavily loaded vehicles which frequently pull manifold vacuum

Metering rods are easily accessible and changing them is quite easy. They are located directly beneath the two airhorn cover plates (arrows), each retained by only one screw.

down to the 8.5-12.5in/Hg range for extended periods. Since these are not maximum load conditions, full enrichment is not warranted and wasteful of fuel.

However, the lean jetting that is a practical companion to a three-step model, would cause surging and stumbling under moderate load with a two-step rod. In addition, a mixture that is too rich for optimum light-throttle cruise economy would be required with two-step rod so that smooth engine operation would be possible during part-throttle acceleration and upgrade hauls. In essence, the three-step rod optimizes fuel economy by tailoring air/fuel ratios more precisely to engine requirements. Some carburetor models are also fitted with a step-up cam and lever assembly which mechanically raises the metering rods. This arrangement provides partial enrichment in direct proportion to the degree of primary throttle valve opening. Full enrichment is still under vacuum control. Such precision is generally not required on passenger cars (with the possible exception of those towing trailers) since vehicle weight is not sufficient to pull manifold vacuum into the area where partial enrichment is required for extended periods of time. Exceptions to this rule are unfortunately becoming more frequent as emissions and fuel economy regulations become more stringent.

ACCELERATOR PUMP CIRCUIT

If the same creative sole who contrived the "economizer valve" label had been allowed to run rampant, he would have no doubt blessed the accelerator pump with an equally obtuse name. Fortunately, after the economizer episode, he was trundled off somewhere far removed from carburetor nomenclature, and spent his remaining years in a small bleak cubicle attempting to balance long columns of debits and credits. As a result, an accelerator pump is an appropriately named item designed specifically to pump fuel into the engine during acceleration. And this remarkable feat of linguistic logic is universally held, except in Great Britain where English is considered a foreign language.

An accelerator pump serves to provide a quick shot of fuel anytime that the throttle is opened abruptly. The volume of fuel discharged through the pump nozzle or "squirter" into the air stream is determined by the amount and quickness of throttle shaft opening. The greater amount of opening and the faster the opening rate, the more fuel will be discharged. If the throttle is opened slowly, fuel is allowed to return to the float bowl rather than being pumped out of the discharge nozzle. And even though this functions similarly to a "leaky" pump, the leakage is controlled and is precisely the arrangement that's desired.

When a standard or mild performance engine is idling, manifold vacuum will generally be between 14 and 17in/Hg. This high level of vacuum will

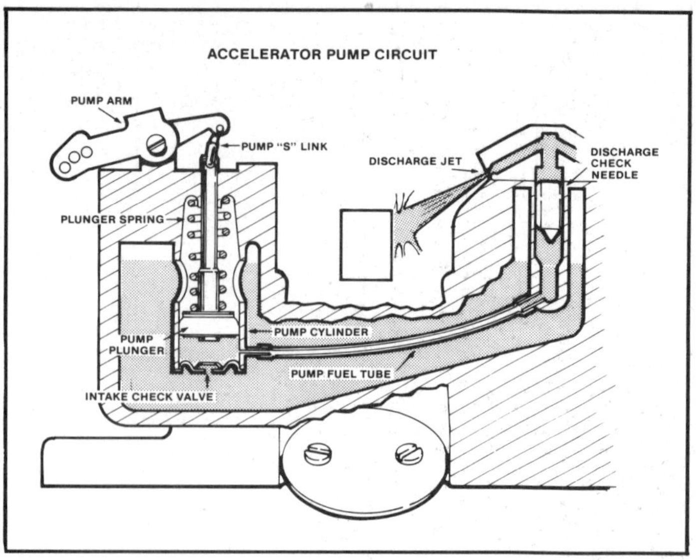

The accelerator pump provides a "quick fix" to the irregularities in air/fuel ratio during the transition from part to full throttle. As the throttle is opened, a plunger is driven down in a cylinder to force fuel through an orifice (discharge nozzle) into the airstream. The fuel is usually aimed at the booster venturi to help breakup the liquid into droplets.

maintain a satisfactory fuel flow through the idle circuit; the air/fuel ratio is normally 13:1. If the throttle plates are opened quickly (as in normal acceleration) several events, which serve to disrupt the steady mixing of air and fuel, occur concurrently:

A) manifold vacuum drops to zero, so fuel flow through the main discharge nozzle virtually stops

B) fuel already in suspension returns to a "more-liquid" state and falls out of the air stream

C) the new throttle setting requires a significant increase in fuel flow so the main system and/or power enrichment circuit must be activated. However without a reasonably strong manifold-vacuum signal, flow will not be initiated.

BRIDGING THE GAP

Without an accelerator pump to supply additional fuel during throttle opening, momentary fuel starvation would cause the engine to sputter, backfire, or die. In essence, this circuit fills a gap in the fuel delivery capability of carburetors used in almost all automotive applications. It should be noted, however, that some low air-flow capacity carburetors—designed for small displacement engines—do not have accelerator pumps. In these special applications, large booster venturis, combined with small throttle bore diameters, are capable of maintaining sufficient vacuum to promptly activate the main system and prevent a momentary lean-out. But almost all carburetors designed for performance or racing—because of their large throttle bore diameters—use accelerator pump circuits.

During the transition from idle flow to main

All Carter four-barrels utilize piston type accelerator pumps. The plunger operates in a cylinder (indicated in photo) that draws fuel from the left-hand float bowl and pumps it into the discharge nozzle located in between the primary venturis. This method is simple, effective, and easy to repair when age catches up with the leather or rubber plunger cup.

This accelerator pump piston is typical of the type used by Carter. The override spring on the shaft prevents damage to the linkage should the throttle be opened faster than the pump can function. Another smaller spring located in the cylinder is used to load the piston cup against the cylinder wall.

Fuel enters the pump cylinder through this one-way check valve located in the fuel bowl floor.

When the pump piston is pushed downward, it moves fuel through passage in the chamber floor that feeds the discharge nozzle. Nozzle size can be modified to alter fuel discharge rate—the larger the nozzle, the faster the fuel squirt. More volume is *NOT* injected when the nozzle is enlarged—this is only accomplished by increasing the pump stroke.

Beneath the nozzle may be found a check needle which prevents air from being drawn into the pump cylinder as it fills with fuel while readying for the next stroke. *Don't leave this out during a rebuild.*

system metering, fuel discharged through the accelerator pump compensates for the *time lag that exists between demand (throttle opening) and main system start up*. This lag is present only when the throttle is opened quickly. During extremely slow opening, air flow increases at a sufficient rate to allow adequate transition from idle to main metering, obviating the requirement for additional fuel.

Alteration of the pump-discharge-nozzle diameter, pump capacity or pump-actuation lever geometry influences both the volume and timing of the "squirt" (how much, how early/late, and how long). Carter carburetors use an external rod and lever arrangement for accelerator pump activation. Tailoring of the fuel delivery through the pump circuit is accomplished by bending the connecting rod or reinstalling it in a different hole in the pump arm.

ENGING LOAD & THE PUMP CIRCUIT

One important factor that is not often considered in the tailoring of the accelerator pump circuit, is engine load. Many times an engine will accelerate smoothly when the transmission is in neutral, but will stumble badly when called upon to propel the vehicle. Absence of a load enables engine RPM to increase quickly when the throttle is opened; manifold vacuum therefore returns to or exceeds the level achieved during idle without much delay. The combination of high air flow, a strong vacuum signal created by the absence of load, and a rapid increase in RPM initiates main system fuel flow quickly. Conversely, when a load is present, air flow and vacuum remain low for a considerably greater time period, during which the carburetor may not be flowing a sufficient amount of fuel to maintain smooth operation.

The duration of the low vacuum signal time period during vehicle acceleration is influenced by many factors: engine torque, vehicle gearing, vehicle weight, ignition timing, cam timing, carburetor size, and the engine RPM range. For these reasons tuning the accelerator pump circuit is often necessary for optimum engine operation, for as low-signal (or lag) time increases, the volume of fuel discharged at the pump nozzle and the duration of the pump shot should also be increased. In essence, the fuel delivered by the accelerator pump must keep the engine supplied with the correct air/fuel ratio mix during throttle transition times, when the fuel available from other carburetor circuits is inadequate.

PISTON VS. DIAPHRAGM

Most carburetor manufacturers use either a piston or diaphragm acceleration pump design. The latter type is generally considered a more efficient pump since fuel leakage out of the pump chamber back into the fuel bowl is virtually non-existent. Carter universally uses a piston type pump and, although it does not offer a positive seal, the "leakage" can slightly improve fuel economy. When the throttle is opened slowly and additional fuel is not required, the piston cup is not wedged against the pump wall and, as the piston moves through its stroke, fuel is not delivered to the nozzle. The inlet valves used in diaphragm-type pumps tend to seat very quickly and consequently this type of pump may discharge small amounts of fuel unnecessarily. While the differences between the two types of pumps are measurable, the effect on overall performance/mileage is generally insignificant.

A disadvantage in the piston-pump design is that the cup (or plunger) can shrivel and harden with the passage of time, allowing more fuel to leak by the piston, and consequently reducing the fuel discharged at the nozzle. In this case, old carburetors never die, they're shipped to Wrinkle City.

Under normal circumstances, when the accelerator pump is actuated, rotation of the throttle lever raises the link rod and one end of the pump arm. This action moves the other end of the arm downward, forcing the piston down—against a compression spring that is part of the piston assembly and an additional spring that aids in sealing the plunger cup—and discharges fuel from the pump chamber. From the pump chamber, fuel travels through a connecting passage and finally exits the circuit at the pump discharge nozzle. A check needle valve, which seats in the main body just below the nozzle, prevents air flowing by the nozzle from drawing more fuel out of the pump circuit after the pump has delivered its full squirt. When the throttle is returned to the idle position, the piston is lifted to its "at rest" position, causing the chamber to refill with fuel. Link geometry and location on the pump arm determine the amount of fuel discharged per stroke. (Additional information pertaining to pump circuit modifications may be found in Chapter 5.)

IDLE SYSTEM

When applied to a piece of machinery, or the chronically unemployed, (or sometimes to individuals who are inactively employed) the term "idle" refers to a rest or non-operating condition. However, when an internal combustion engine "idles," it is not at rest, but merely running at the slowest speed in its usable operating range. Some race engines idle at RPM levels that are in excess of the cruise speeds of stock engines, so idle is really a relative term, and "acceptable idle speed" is largely dependent upon the application for which an individual engine is built and tuned.

To people born on the "other side" of the generation gap, or this side of the ecology fad, engine idle is acceptable only if it is smooth and slow, a condition that is primarily dependent upon camshaft profile. Performance enthusiasts will typically accept (and usually relish) some amount of roughness at idle as a trade-off for increased high RPM horsepower. Race car owners will accept all the idle roughness they can get, if it's part of the trade for more power higher up the RPM scale. With these parameters, it is easy to understand why idle circuit calibrations differ drastically between carburetors. It should also be obvious that modifications to the idle system will almost always be required when a carburetor is used in an application for which it was not originally intended.

Dilution of the intake charge by exhaust gases that have not been purged from the combustion chamber is one of the phenomenons of the internal combustion engine, and it is particularly exacerbated by engines equipped with high-performance camshafts. This condition is most severe at low engine speeds, but dissipates as engine RPM approaches the point of optimum operating efficiency. Low-speed dilution necessitates that idle mixtures be richer than cruise air/fuel ratios in order to maintain some degree of smoothness. Typically, as camshaft duration and overlap are increased (i.e., the cam profile becomes more radical) idle fuel jetting (idle feed restriction size) must be correspondingly increased to compensate for mixture dilution and over-scavenging (an additional ill effect of high-performance cam profiles operated at low speeds, resulting in an amount of the unburned intake charge being drawn out through a late-closing exhaust valve). It is for this reason that carburetors designed for race and high-performance engines frequently contain an idle fuel calibration that is too rich for a street engine.

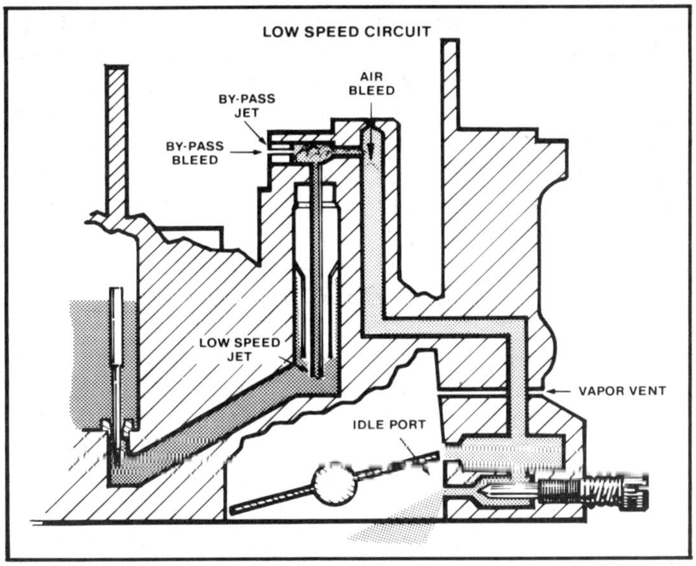

The idle circuit may be thought of as a main metering system for low-speed use. The exact design may vary, but fuel is almost always mixed with air in a two-step sequence; first in the emulsification circuit, then in the manifold after it is discharged from the carburetor.

IDLE CIRCUIT FUNCTION AND CALIBRATION

Given the wide variety of applications in which carburetors of equivalent performance potential are employed, it is understandable that idle calibrations can differ significantly between part numbers. For example, a 625cfm "Competition Series" AFB and a 9000 model carburetor of the same airflow capacity are calibrated for different idle mixtures, each providing optimum idle function for its designated application.

In function, the idle circuit is very similar to the main metering system and may be thought of simply as a main system for low-speed use. Under normal circumstances, idle flow begins by engine vacuum acting on the curb idle discharge port (the feed hole located below the throttle plates) and creating a fuel delivery signal much like the one that activates the main system at high air flow rates. Vacuum pulls fuel out of the main well through an idle feed restriction, and then through a series of passages where it is ultimately emulsified with air admitted at the idle air bleed. Once emulsified, the air/fuel mixture is drawn past the mixture adjusting needle and out of the idle discharge port. The routing of idle fuel and fuel emulsion differs between carburetor models, but conceptually, the idle circuit is virtually identical from one carburetor type to another. In most cases, idle passages lead fuel up toward the top of the carburetor, where bleed air is admitted, before bringing it back down to the discharge port. This is done in order to raise idle fuel above float-bowl level. Without this up-over-and-down routing, the discharge port would serve as a float bowl drain, siphoning fuel from the float bowl into the manifold.

Traditional designs have idle mixture adjusting screws protruding into the discharge ports. Some carburetors, however, represent a departure from this practice since the mixture screws are located at some distance from the actual port. In operation, the two designs are identical. The adjusting needle simply controls the amount of fuel emulsion which is allowed to reach the port. Placement of the needle is actually of little consequence.

MAKING THE ADJUSTMENT

Contrary to popular belief, the mixture needle *does not* control idle-circuit air/fuel ratio. Although this may fall under the topic of semantics, there is a difference between the amount of air (emulsification) in the idle-circuit mixture, and the mixture that ultimately reaches the engine. Internal air/fuel idle circuit calibration is determined by the diameters of the idle feed restriction and idle air bleed. It should be remembered that an "idle restriction" is nothing more than a metering jet for the idle system, and that the air bleed serves as a primary air flow regulating orifice. As the air/fuel emulsion exits the idle port, it is mixed with air that flows past the throttle plates. Varying the throttle plate opening (with the idle speed screw) increases or decreases both the amount of air that is allowed to flow into the intake manifold, and the volume of air/fuel emulsion discharged at the idle port. Since these two idle mixture components are both affected by the degree of throttle opening, the ratio of air to air/fuel, once set, is not drastically altered by changes in idle speed. Turning the idle mixture screw varies the volume of air/fuel emulsion discharged into the manifold, *not* the ratio of air to fuel in the emulsion. Therefore, when a satisfactory idle cannot be achieved by "tweaking" the mixture screws richer (more air/fuel to air) or leaner (less), it can be assumed that the carburetor idle calibration is not within the range demanded by that particular engine.

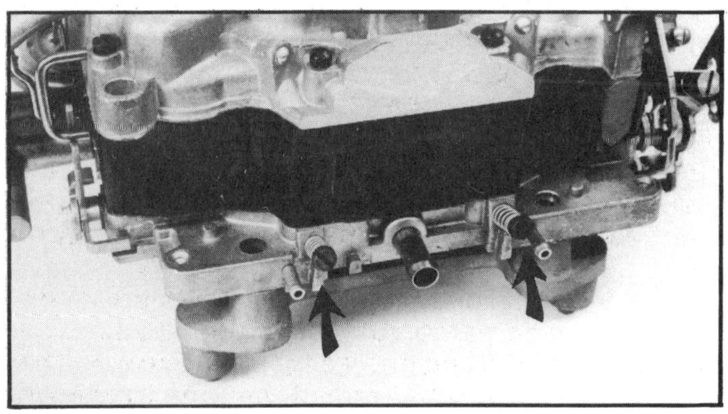

After being mixed with air, fuel in the idle circuit flows through a passage that leads it by a tapered needle attached to the adjusting screw. Rotating the screw counter clockwise backs the needle out, allowing a greater volume of air/fuel mix to flow into the throttle bore. Turning the screw clockwise has the opposite effect—usually. There is one exception known as the "reverse idle" system where only air is metered by the adjusting screws.

Most carburetors, like this Thermo-Quad, have two idle speed adjustments. The lower screw (large arrow), which controls high speed idle, contacts a cam when the choke is in use. Once the engine warms up, the cam moves out of position, and idle speed is controlled by the normal idle speed screw (small arrow).

Use of dual four-barrel carburetors is often an excellent example of idle circuit incompatibility. With such an installation, there are two primary idle systems and the volume of air/fuel emulsion is effectively double that available from a single four-barrel mounted on the same engine. Even when the mixture screws are adjusted toward lean, the requirement for uniform fuel distribution may simply cause too much overall fuel flow to permit a clean idle. In such cases, the idle-feed restrictions of both carburetors should be reduced (in cross-sectional area), to lean the idle mixture to an acceptable level.

Idle mixture calibrations are rather crucial, especially in these days of stringent emissions regulations. Since the original settings rarely need alterations for standard applications, most manufacturers supply carburetors with fixed metering restrictions; either eliminating the idle adjusting screws altogether, or severly limiting there adjustment range. However, when idle system calibration changes are required, the restrictions must either be drilled larger or removed and replaced with ones containing a smaller diameter orifice. These types of alterations are confined almost exclusively to highly specialized race applications, where precision equipment and dyno facilities are used to verify the results of all modifications. Fortunately for the "average man," satisfactory idle performance can, for the most part, be achieved by varying the settings of the idle mixture screws and the idle speed control screw.

In most instances, as little as an eighth of a turn will alter idle quality. In addition to adjusting for either the highest manifold vacuum or engine speed, the left and right needles should be balanced as closely as possible (usually done by backing out the needles the same number of turns).

TURN, TURN, TURN, BUT WHICH WAY

Until a few years ago, rotating an idle mixture screw (on just about any carburetor) counterclockwise always meant that more air/fuel mix would be discharged and a richer idle mixture would result. But when the bureaucrats in Washington began increasing their stranglehold on Detroit, carburetor manufacturers found that a new system offered advantages for meeting emissions standards. Some companies incorporated a second air bleed into the idle system, which was adjustable. By changing a few passages, the mixture screw could then be used to adjust bleed air, rather than fuel. Carburetors incorporating this system are commonly referred to as "reverse idle" models since the engine receives a richer mixture when the screws are turned clockwise, and a leaner mixture when they're rotated

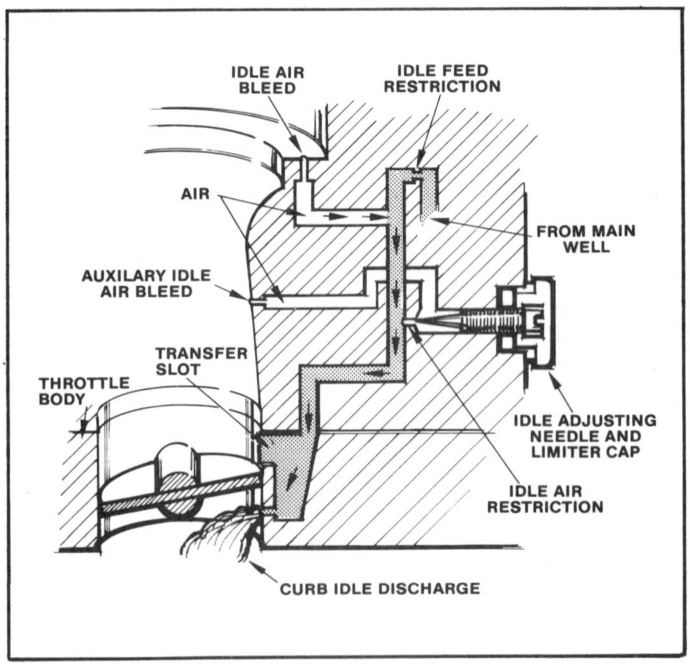

The reverse-idle system is actually a variable air bleed arrangement. Idle fuel mixture is controlled (richer or leaner) by variations in the amount of bleed air admitted to emulisfy the idle fuel.

Unless an engine is fitted with a long duration cam, this will be the approximate position of the throttle plates at idle. In some instances, up to .060-inch of the transfer slot (not shown in this photo) may be exposed without seriously impairing idle quality.

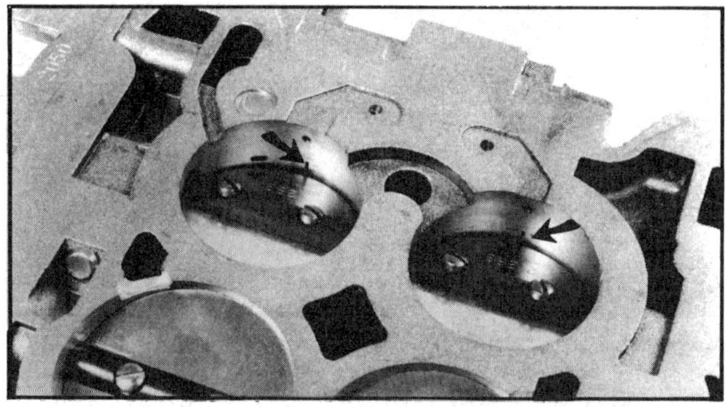

Idle requirements with racing cams sometimes necessitate a wide throttle plate opening that exposes a large portion of the transfer slots. Since fuel flow through the slots is not controlled by the mixture adjustment screws, the screws cease to have an influence on idle quality.

counter-clockwise. Identification of a "reverse idle" carburetor is usually facilitated by a special tag containing an arrow indicating the direction that the mixture screw should be turned to obtain a leaner mixture. And to further complicate the issue, some "reverse idle" carburetors are equipped with left-hand threaded adjusting screws, so they still may be adjusted by the "clockwise lean, counter-clockwise rich" procedure.

Implementation of the new, reverse-idle system is actually quite simple. But first, some basics.

In a conventional carburetor, separate passages are used to supply air/fuel mix from the idle circuit to each idle discharge port/transfer slot—the transfer slot is an additional idle-fuel outlet located above the normal idle discharge that is activated as the throttle plates are opened slightly, eliminating "off-idle stumble." (Additional information on the function of the transfer slot is provided in an upcoming section.) As fuel flows up from the idle well, across a connecting passage into the idle-down passages, it passes the transfer slots and continues down to the idle discharge ports, through which the needle end of the mixture adjusting screws protrude. Normally, as the mixture screw is rotated counter-clockwise, the fuel delivery passage is opened wider and more air/fuel mix flows from the idle discharge port. Turning the screw clockwise has the reverse effect, and when the needle is firmly seated against its seat, no fuel reaches the idle port.

In a "reverse-idle" carburetor, the mixture screw opens or closes an auxiliary air bleed, rather than protruding through the idle port and regulating fuel flow. Since the idle discharge ports are supplied by the same channel that feeds the transfer slots, inclusion of an auxiliary air bleed means that the air/fuel mixture discharged through both the idle port *and* the transfer slot will be regulated by the mixture screw. This occurs because the adjustable air bleed controls the amount of air admitted into the entire idle circuit, not just at the idle port. Fuel flow out of just the idle port is therefore no longer adjustable, and volume is strictly the result of idle-feed restriction calibration. This is why it is almost impossible to kill the engine when adjusting the idle mixture on a "reverse idle" carburetor. Regardless of the screw setting the engine will be supplied with sufficient idle fuel.

As previously mentioned, with a conventional idle circuit, the mixture adjusting screw is always downstream of the transfer slot, and the amount of fuel discharged through the transfer slot is never affected by mixture needle position. However, on a "reverse idle" system, mixture screw adjustment does change the air/fuel ratio of the mixture drawn out of the transfer slot and this can have a significant impact on gas mileage since the slot supplies much of the fuel used during "off-idle" highway cruise.

On Carter carburetors equipped with an auxiliary adjustable idle air bleed, a single adjusting screw is used to control both right and left side circuits. Standard mixture screws are also contained in these carburetors, however, they are recessed in the carburetor body, sealed and unaccessible. The air-adjusting screw is machined with left-hand threads, so mixture adjustment may be made according to standard "clockwise-lean" procedures.

Obviously, in the common 12:1 to 14:1 air/fuel ratio range, the proportion of air to fuel is rather lopsided. Not that it shouldn't be that way, considering the cost of each, but the disparity of amounts calls for some degree of consideration when circuit operation is altered; such is the case in the "reverse idle" system. The tip of the air-adjusting screw had to be recontoured so that more adjustment was obtainable per turn of the screw. Whereas the end of a standard idle mixture screw is machined into a long tapered needle, an air bleed screw is more like a blunt wedge. Were the needle configuration used to control bleed air, the screw would have to be rotated several turns in order to effect any appreciable change in mixture.

A FREE TRANSFER

Just as the term "economizer valve" is a mis-

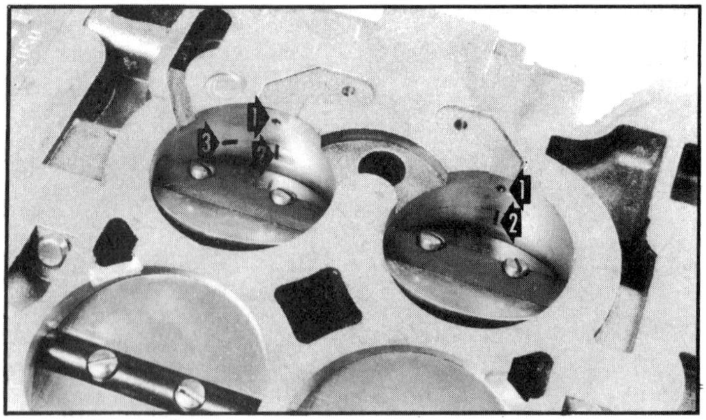

As the throttle is opened a point is reached where fuel flow through the curb idle discharge ports (arrows 1) is inadequate yet air flow is not sufficient to initiate flow through the main metering circuit. The transfer slots (arrows 2) provide a smooth transition from idle to main circuit flow. The horizontal slot (arrow 3) is a source of ported vacuum.

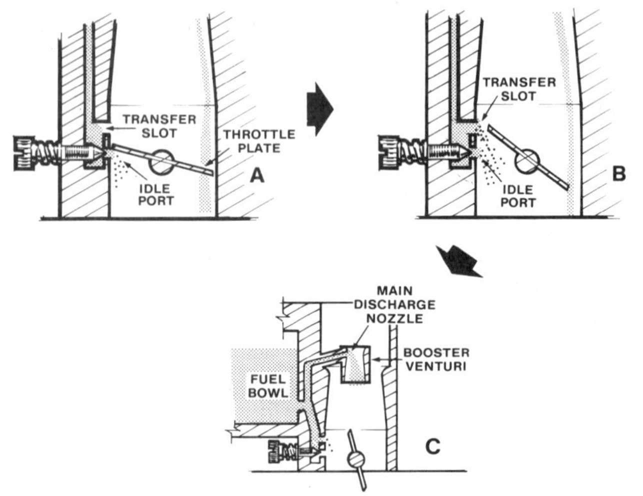

These three drawings indicate fuel flow and transition through the idle, transfer and main circuits. (A) At closed-throttle idle, the transfer slot is above the throttle plate and "sees" only atmospheric pressure, therefore only the idle port discharges fuel. (B) As the throttle is opened and the transfer slot exposed to manifold vacuum, additional fuel is drawn into the air stream. (C) Under full (or nearly full) throttle, there is insufficient pressure drop across the idle or transfer slots for fuel to flow, however, air flow across the booster venturi has reached a sufficient level to initiate main-system fuel flow. A properly designed carburetor will smoothly transition from one circuit to the next, shutting one down only after the next has taken over; the result is a uniform mixture of air and fuel at any throttle setting.

nomer, "idle system" is somewhat anomalous. The circuit does function at idle, but also feeds the transfer slot which serves to bring an engine off idle and into a higher RPM range. (Not wishing to commit another nomenclature faux pas, I won't attempt to rename the circuit). As the throttle plates are opened, the portion of the transfer slot exposed to manifold vacuum increases, bringing about a corresponding increase in fuel flow. Eventually, the throttle plates will be opened sufficiently to actuate the main system, and flow through the idle circuit will diminish when the main system fuel demands "starve" the idle circuit. The purpose of the transfer slot is as its name implies, to smoothly transfer fuel control from the low-speed idle system to the high-speed main system.

Although there is no provision to "shut-off" idle fuel once the main system is activated, this is effectively what occurs. Once the throttle plates are open far enough to allow passage of a significant quantity of air, the amount of vacuum presented to the idle port and transfer slot is reduced. Concurrently, a strong signal has been developed at the main discharge nozzle, so fuel that might otherwise reach the idle system is "stolen" from the main well by the main circuit. (Both the idle and main systems are supplied by the main well.) Since idle fuel is neither in demand nor in supply, flow effectively ceases.

CHOKES AND COLD STARTING

When the cold of night (or day for that matter) strikes, it isn't just we mere mortals that are brought discomfort. Our friend and ally, the internal combustion engine is right beside us shaking and shuddering, albeit for quite different reasons. It is a simple fact of automotive life that gasoline does not readily vaporize when both it and the metallic passageways through which it must flow, are very cold. Add to these considerations the comparatively low speed (and attendant low vacuum) at which a starter spins an engine, and it can be seen that cold starting does not provide ideal conditions for engine operation. Without a good strong signal, very little fuel is drawn out of the carburetor, and most of what flows returns to liquid state as soon as it contacts a cold manifold runner.

A choke plate is the device used to alleviate problems associated with coaxing a cold engine to life. When the plate is closed, a vacuum is created immediately below it (even by a slow turning engine) and fuel is drawn out of *both* the idle *and* main system. With vaporization assisted by an over abundance of fuel, conditions are more conducive to combustion and an engine will usually spring to life. Once this occurs, there is sufficient vacuum to initiate normal fuel flow and cause a reasonable amount of vaporization. Fuel will still condense on the intake manifold walls, but a richer than normal mixture is continually supplied until the manifold warms up and the choke is fully open.

The choke plate functions in conjunction with a fast idle cam that raises engine speed when the choke is in operation. Most automatic choke systems are also fitted with a "pull off" diaphragm which partially opens the choke when manifold vacuum is developed (i.e., when the engine starts); preventing an over-rich mixture and subsequent stalling.

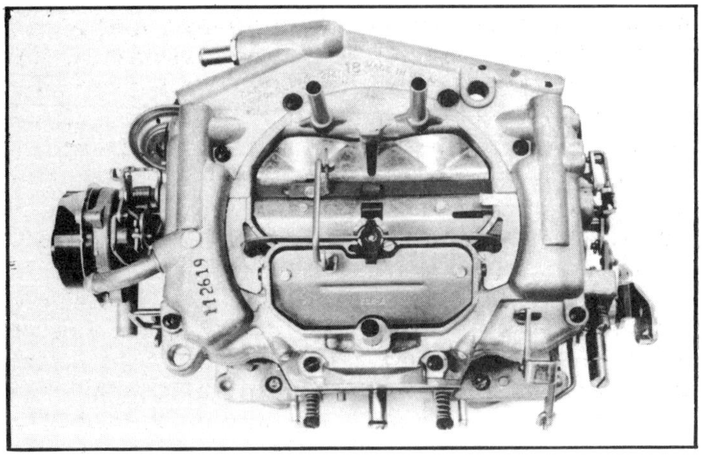

When a choke plate is closed and the engine is rotated by the starter, a low pressure area is created in the venturis. Fuel then flows out of both the main discharge nozzle and idle ports, providing the rich mixture necessary in cold starting.

The choke linkage on Carter four-barrels is somewhat intricate because, in addition to typical functions, a secondary throttle lock-out is employed that prevents the secondaries from opening while the engine is cold. This view shows the lockout (1), fast idle (2) and side-to-side transfer linkages (3).

INTEGRAL, DIVORCED, ELECTRIC— SOLUTIONS TO A CHOKING SITUATION

Several types of choke systems have been used over the years. Older carburetors were equipped with an integral choke, wherein hot air (sometimes from the exhaust manifold) was piped directly to a bimetal thermostatic spring, which warmed, changed its shape, and opened the choke. Vacuum was also applied to the choke housing so that the hot air was actually drawn into the spring chamber. This type of choke can be problematical during hot starts because the thermostatic spring will cool considerably faster than the engine, allowing the choke to close even though the engine is still warm enough

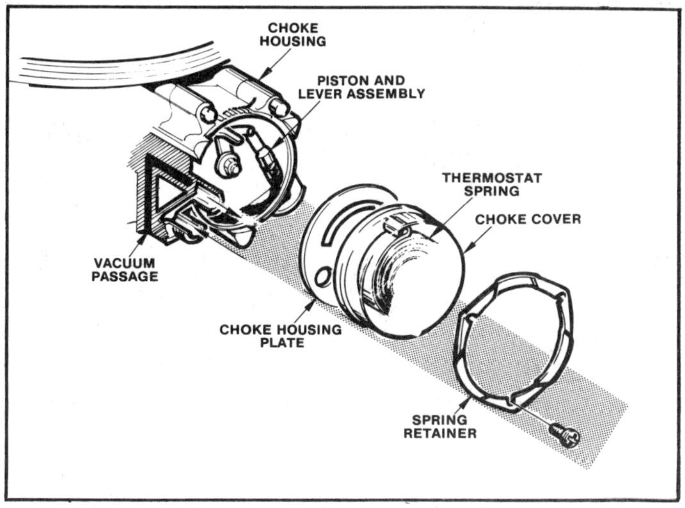

Integral (and some electric) chokes use manifold vacuum to draw hot air through the choke housing to activate a thermostat spring that, in turn, counterbalances a piston and lever assembly acted upon by manifold vacuum. This system closes the choke when air (usually drawn in from the exhaust manifold area) is cold and slowly opens the choke flap as the engine warms.

for an open-choke start. Deterioration with age is another problem. Carbon build-up in the hot air tube can reduce or completely stop the flow of warm air to the spring. This results in a choke that either opens very slowly, or not at all.

An alternative to the integral choke is the divorced type, wherein the thermostatic spring is mounted either in a well or on a pad on the intake manifold, just above the exhaust crossover passage. A rod connects the bimetal spring to the choke control lever, and a "pull-off" diaphragm is used much in the manner that it is employed on an integral choke.

Electric chokes have achieved a great deal of popularity because they alleviate the need to match choke actuating mechanisms to specific engines. An electric choke functions in exactly the same manner as an integral type but instead of using exhaust gasses to heat the bimetal spring, an electric heating coil is employed.

There's a Carter carburetor in there somewhere! Several late model "emissions type" powerplants, particularly Chrysler engines, use the Carter carburetors. The normally limited use of the secondaries on these engines is well suited to the common primary/secondary float bowl design (no secondary idle circuit is required to keep fresh fuel in the bowl), and the simplicity of the Carter design allows lower initial cost and easier maintenance.

400-CID 4-BARREL ENGINE

Engines produced during the Sixties and early Seventies frequently made use of a divorced choke. With this style, the bimetal spring is located in a well in the intake manifold. Heat from the exhaust cross-over passage warmed the well and consequently the spring, causing the choke to open.

A primary difference between a divorced and integral type of choke is the divorced unit's lack of adjustability. The phenolic housing used with the integral choke is typically marked "Rich" and "Lean." Rotating the housing in the "Rich" direction increases the spring preload, causing the choke plate to open more slowly. Turning the housing toward "Lean" logically has the opposite effect. Some divoced choke bimetal units had a provision for similar adjustment, but this usually required removing the spring assembly from the intake manifold and was largely abandoned on later installations.

Use of electric chokes was instituted as a means of allowing aftermarket replacement carburetors to fit a wide variety of intake manifolds. An electric choke is actually an integral type except that the spring housing contains a heating element that is connected to a 12-volt power source controlled by the ignition switch. When the car is started, and current is applied, electrically induced heat causes the same type of bimetal reaction as hot air being piped in. The electric choke allows for a great deal of versatility on custom carburetor/intake manifold installations. However, it should be noted that power for the choke should not be drawn from the coil or other ignition components, since the current drain can adversely effect ignition system performance.

In days past, a hand choke was the only alternative to extremely difficult cold starting. The beauty of a hand choke is its simplicity; a cable with a knob attached to one end controls operation. It is the person behind the wheel with whom choke control resides and he can decide whether a quick or slow opening is appropriate. He can also forget to open the choke as the engine warms, and if he doesn't notice the black smoke belching from the tail pipe, he will experience an immediate degradation of fuel economy.

In some warmer climes, it is possible to drive year around without a choke of any type. Cold starting is accomplished by depressing the gas pedal several times so that the accelerator pump squirts a healthy volume of raw fuel into the intake manifold. This will enable the engine to start, but it may run very erratically until it warms enough to allow normal vaporization.

SECONDARY METERING SYSTEMS

In Carter four-barrel carburetors, the secondary metering circuits differ significantly from those used on the primary side. The differences exist because of a variation in operational parameters. The secondary barrels are used only a fraction of the time, (except by over enthusiastic "stab-it-and-steer-it" types) therefore special features must be included to accommodate comparatively rare and intermittent use.

THE SECONDARY IDLE CIRCUT

A prime example of this special design is the secondary idle system, or lack of it. Of all Carter four-barrels, only the WCFB is fitted with a secondary idle

The WCFB, with individual float bowls for primary and secondary circuits, is the only Carter carb equipped with secondary idle circuits. They aren't required in the other models because each fuel bowl feeds both a primary and secondary circuit, preventing fuel stagnation when the secondaries are not used for long periods of time.

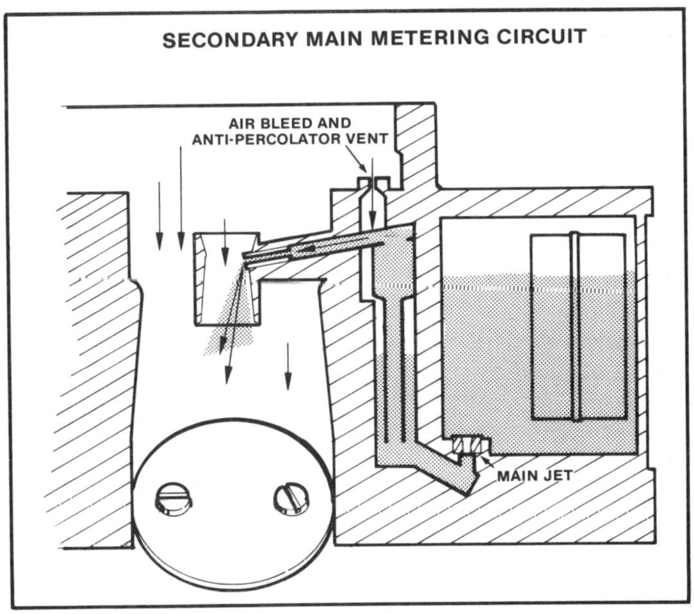

Since the fuel bowls for the secondaries on the AFB, AVS, and the Thermo-Quad are shared with the primaries there is no need for a secondary idle circuit (as illustrated above). However, the WCFB uses a separate float bowl for the secondaries and must "bleed" fuel from this reservoir with a secondary-idle circuit to prevent stagnation during long period of non use.

Although the secondary main metering circuitry is very similar from model to model, the discharge nozzle configuration varies. WCFBs and AFBs have nozzles located in the secondary booster venturis. However, the AVS and Thermo-Quad (the TQ airhorn assembly is shown here) have no secondary venturis so the discharge "nozzles" are simply perforated tubes which protrude into the air stream.

Theory says that two AFBs actuated by mechanical linkage will cause an engine to stumble momentarily when the secondary throttles are opened. However, the AFB has a counterweighted air valve that smooths the transition and eliminates hesitation on this 426 Street Hemi.

circuit. Whereas the AFB, AVS and Thermo-Quad utilize left and right side fuel bowls, the WCFB float chambers are separated on a primary/secondary basis. If no fuel was taken from the secondary side at idle, and if a driver was overly conservative in his application of the accelerator pedal, fuel in the secondary bowl would be allowed to stagnate. After a prolonged period of non-use, varnish and gum would begin to block various passages, and when the secondaries were finally called into action, fuel delivery would be inadequate or totally blocked. By including a secondary idle circuit, (on carburetors with separate primary and secondary float bowls) fuel continuously flows through the float bowl and out the constant feed discharge hole (the secondary equivalent of the curb idle discharge port). This insures that the secondary fuel bowl is always filled with fresh fuel, eliminating varnish build-up. Secondary idle fuel can also help improve fuel distribution at low engine speeds.

On all the WCFB and other carburetors with a secondary idle circuit, it is supportive, rather than being essential to proper carburetor/engine operation, and it usually has no provision for mixture adjustment. An exception to this may be found on certain (non-Carter) race carburetors where adjustment provision is included to ease tuning of engines with long duration camshafts. Except for the fact that the secondary idle system passes less fuel and is not (generally) adjustable, it is identical to the primary version.

SECONDARY MAIN METERING

The secondary circuits of most four-barrel carburetors are actually quite simple since they are only called into action during periods of high load (i.e., WOT—wide open throttle—or "jack-rabbit" starts as aging relatives are fond of calling them). Like most carbs, Carter four-barrels have relatively rich secondary fuel calibrations and no provision for auxiliary

With the secondary boosters removed, the AFBs secondary air valve is exposed. It is sometimes left out by carb "tuners" in search of improved acceleration, but in most instances, it is quickly replaced as a very noticeable "bog" will accompany off-the-line acceleration.

On AVS carbs, a spring-loaded rather than counterweighted air valve is used. It is located above the discharge nozzles. The opening rate may be easily adjusted by altering the spring-tension adjuster (arrow).

enrichment (in the form of an economizer, etc.). Air/fuel ratios are established by jet size and remain constant regardless of manifold vacuum. (It should be noted that a few carb models do utilize power enrichment circuits on the secondary side. However, these are the exception rather than the rule.)

VACUUM SECONDARY ACTUATION

If the secondary throttle plates were linked directly to the accelerator pedal, the profits reported by the oil companies would be even greater. (If that's possible.) Constant use of all four barrels is not only wasteful, it would keep most engines in a state of over-carburetion. In order to correlate the degree of secondary throttle opening to actual engine requirements, carburetor manufacturers use one of several forms of venturi-vacuum activation. The method used commonly by Carter is a counterweighted or spring-loaded air valve. Irrespective of the method employed, the concept is designed to precisely tailor carburetor air-flow capacity to engine requirements, rather than providing "too much too soon," as can be the case with mechanical secondary opening.

AIR VALVE SECONDARIES

In the air valve controlled secondary, as employed by Carter, manifold vacuum and air velocity (rather than venturi vacuum) are the primary factors that control secondary actuation. The air valve, sometimes called an auxiliary throttle valve, is located above the throttle plates and held closed by either a spring or counterweight. Secondary operation becomes a two-stage process, requiring that the secondary throttles be mechanically opened in order to initiate air valve opening; in order for maximum air/fuel to flow in the secondaries, both the air valve and the throttle plates must be fully opened.

In order for the air valve concept to function properly, the throttle plates must be opened mechanically as a means of inducing manifold vacuum and air flow through the venturis. Once the throttle is opened, manifold vacuum initially opens the air valve, however, air flow soon becomes the controlling factor, further opening the valve as flow increases. Since changes in intake-manifold velocity (intake pulses, etc.) occur very rapidly, the net effect is a smoothly opening air valve and steady acceleration (assuming, of course, that all other systems are upholding their end of the bargain).

On carburetors with spring-loaded air valves, such as the Carter AVS and Thermo-Quad, opening rate may be adjusted by either increasing or decreasing spring pressure. On models with counter-weighted valves, such as the WCFB and AFB, opening rate is changed through the addition or removal of weight. Adding weight or spring pressure can often improve standing-start performance in cars equipped with automatic transmissions, while a lighter weight or spring load is usually more effective with manual transmission vehicles, particularly in all-out drag-race situations.

It should also be noted that some carburetor models use air valves that are controlled by a vacuum diaphragm. Operation of these "hybrids" is similar to that of standard vacuum-controlled secondaries, except that the diaphragm controls air valve position rather than throttle shaft opening. They are not commonly used in performance applications.

Like the AVS, the Thermo-Quad utilizes a spring-loaded secondary air valve. However, the valve itself is contoured rather than flat. The shape is designed to alter the venturi effect, and maintain uniform air/fuel ratio and distribution, as it opens.

Fuel enters the float bowls through an inlet valve commonly referred to as a "needle-an-seat" assembly. As the fuel level drops, the float lowers, and the needle is moved away from the seat allowing fuel to enter the bowl. When the bowl is full, the float rises, forcing the needle into the seat, shutting off fuel flow.

MECHANICAL SECONDARIES

As the name implies, "mechanical secondaries" operate through a direct (or mechanical) link to the accelerator pedal linkage. A 1:1 direct linkage (where the secondaries open at the same time and at the same rate as the primaries) is used only on a few "race-only" carburetors. More often, mechanical actuation is controlled by a progressive linkage system. The term "progressive" does not pertain to a forward thinking designer, but to the opening rate of the secondaries relative to the primaries. Progressive linkage will usually allow the primary throttle plates to open 40° to 45° before the secondary plates begin to open. This will improve low speed driveability and fuel economy. But because the primaries are already open 40° or more before throttle opening progresses to the secondaries, the opening rate for the secondaries must be very quick so that both primaries and secondaries reach wide-open-throttle at the same time.

Under most operating conditions, rapidly rotating the secondary throttle plates from closed to fully open would result in a severe stumble. On the primary side, an additional "squirt" of fuel from the accelerator pump prevents this situation, and it is solved the same way on almost all mechanical-secondary systems—with a secondary accelerator pump. The two pump circuits are virtually identical. There are some mechanical-secondary carburetors, however, with low air flow capacities and specially designed booster venturis that operate satisfactorily without a secondary accelerator pump. But all high-performance units require a secondary pump circuit to overcome loss of signal during transition from part to full throttle operation.

FUEL INLET CONTROL

Back in the days of the Model-T the accomplished motorist carried a pair of pliers, some bailing wire and a screwdriver. So equipped, he could fix all but the most serious engine problems. Garages performed "complex" repairs using only a slightly more sophisticated assortment of tools—such as hammers! And, often as not, a carburetor "tuneup" was performed by thumping on the float bowl with any available tool, including a hammer if one was available; as the hammerer attempted to free an inlet needle and seat that had become stuck!

As fuel quality, carburetor manufacturing techniques and metallurgy improved, the need for hammers and the like diminished. The float/inlet valve configurations currently employed by carburetor manufacturers typically provide trouble-free operation for tens of thousands of miles. Like the float, the inlet valve (also called a needle and seat) size and shape differs between carburetor models. However, operational characteristics are virtually identical whether the carburetor in question is a Carter, Holley or Rochester. Either a fuel pump or gravity feed causes fuel to flow from the gas tank to

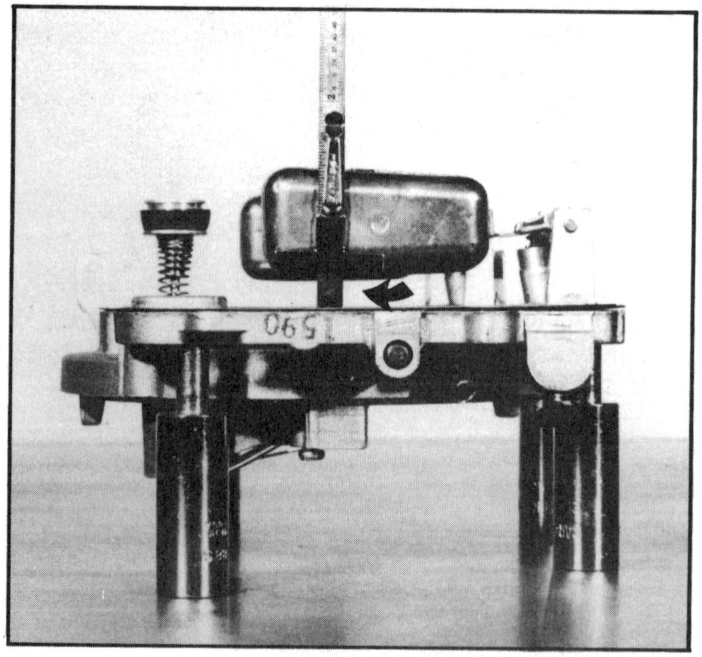

Proper float level is essential to satisfactory carburetor operation. With Carter four-barrels, level is measured by inverting the air horn assembly and noting the distance from the gasket surface to the float body.

the carburetor. Many carbs have a small filter or filtering screen, at the inlet, to remove dirt and other potentially damaging particles. Once past the filter, the fuel flows through the inlet valve into the float bowl. As the volume of fuel in the bowl increases, the float rises and with it goes the inlet needle, often clipped to the tab that extends from the float body to the pivot pin. Ultimately, the float is lifted high enough to force the needle firmly against the seat, thereby stopping the flow of fuel into the chamber.

As the engine operates, fuel is drawn from the float bowl through the main jets, and when enough fuel has taken its leave, the float drops, opening the valve and allowing the level to rise until the needle is once again forced firmly against the seat. Adjustment of fuel level is accomplished by altering the position at which the float causes valve closing. Most

Float drop is another important setting as it influences flow through the inlet needle-and-seat. A bendable tab is often used to control drop, but on the Thermo-Quad float, drop is regulated by the tab stop (arrow-1). The tab stop isn't adjustable, however, float drop can be adjusted by bending the the float stem (arrow-2), then reseting the bendable tab (arrow-3) to regain proper float height.

Carter carburetors utilize large float bowls within the main body. Most have fuel baffles (arrow-1) and all draw fuel through jets either located on or near the bottom of the bowl (arrows-2 indicate primary and secondary jets). Arrow-3 indicates a balancing tube used to maintain consistent right-and left-hand bowl levels.

The Thermo-Quad features a main body of phenolic resin which contains the fuel bowls and primary venturis. This material offers excellent heat insulating properties so fuel in the bowls is kept cooler, increasing its density and power producing potential.

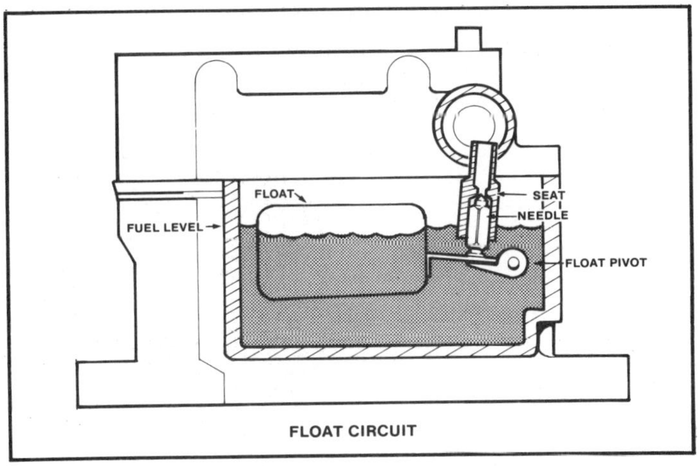

The typical float circuit uses a hollow (or solid nitro phyl) float that, as the fuel level rises, forces a needle assembly into a tapered seat, closing off fuel flow into the bowl. Most Carters use one float bowl on each side of the carburetor (one for each primary/secondary throttle pair). In these configurations, there is a connecting channel to equalize right/left fuel levels.

carburetors (Rochester, Carter and non-modular Holleys) utilize an internally adjustable float. (Only Holley modular carburetors are equipped with externally adjustable floats.) Alteration of fuel level is accomplished by bending the tab which attaches the float to the pivot pin. Forming the tab so that the float is moved downward lowers fuel level (the float "reaches deeper" into the bowl and less fuel, therefore, is required to lift the float and shut off the flow of fuel). Raising the float in relation to the pivot has the opposite effect.

In some carburetors, float drop (how far the float will decend into an empty bowl) is also adjustable. For high-performance use, drop is usually set beyond specifications to insure a "wide-open" path for fuel to flow past the needle. Insufficient float drop will place the needle in the path of incoming fuel, creating a flow restriction even when the needle is in the wide open position.

Proper adjustment of fuel level is critical to satisfactory engine operation. If the level is too low, activation of the main system is delayed, fuel mixtures can become too lean and the engine may temporarily run out of gas. Too high a level can lead to over-richness, flooding and fuel slosh out of the vent tubes.

FLOAT BOWLS

Float bowls, floats, and needle-and-seats vary according to manufacturer's design and the intended application. Some designs have separate removable bowls, but most carburetors—and all Carters—utilize cavities in the main body to hold the fuel supply.

Carter has used several arrangements over the years. The original WCFB was fitted with four fuel reservoirs and a corresponding number of floats. However, the two primary floats were attached to a common lever assembly and used a single inlet needle; the secondary side was handled in a similar fashion, so even though there were four "floats" there were only two "float assemblies."

The AVS, AFB and Thermo-Quad are each equipped with two fuel reservoirs, floats and inlet needles. But rather than being separated on a primary/secondary basis there is a right/left division. Each bowl supplies fuel to one primary and one secondary barrel. There is a connecting channel that equalizes slight differences in float bowl levels.

SETTING OF FUEL LEVEL

Fuel level may be checked and adjusted through a variety of procedures, as determined by carburetor type. Adjustment is usually determined by a measurement of the float distance from a reference surface. This dimension may be checked either by allowing the bowl to fill with fuel (on models where fuel enters through the main body rather than air horn assembly,) or by removing and inverting the air horn.

CHAPTER 5 CARBURETOR MODIFICATIONS

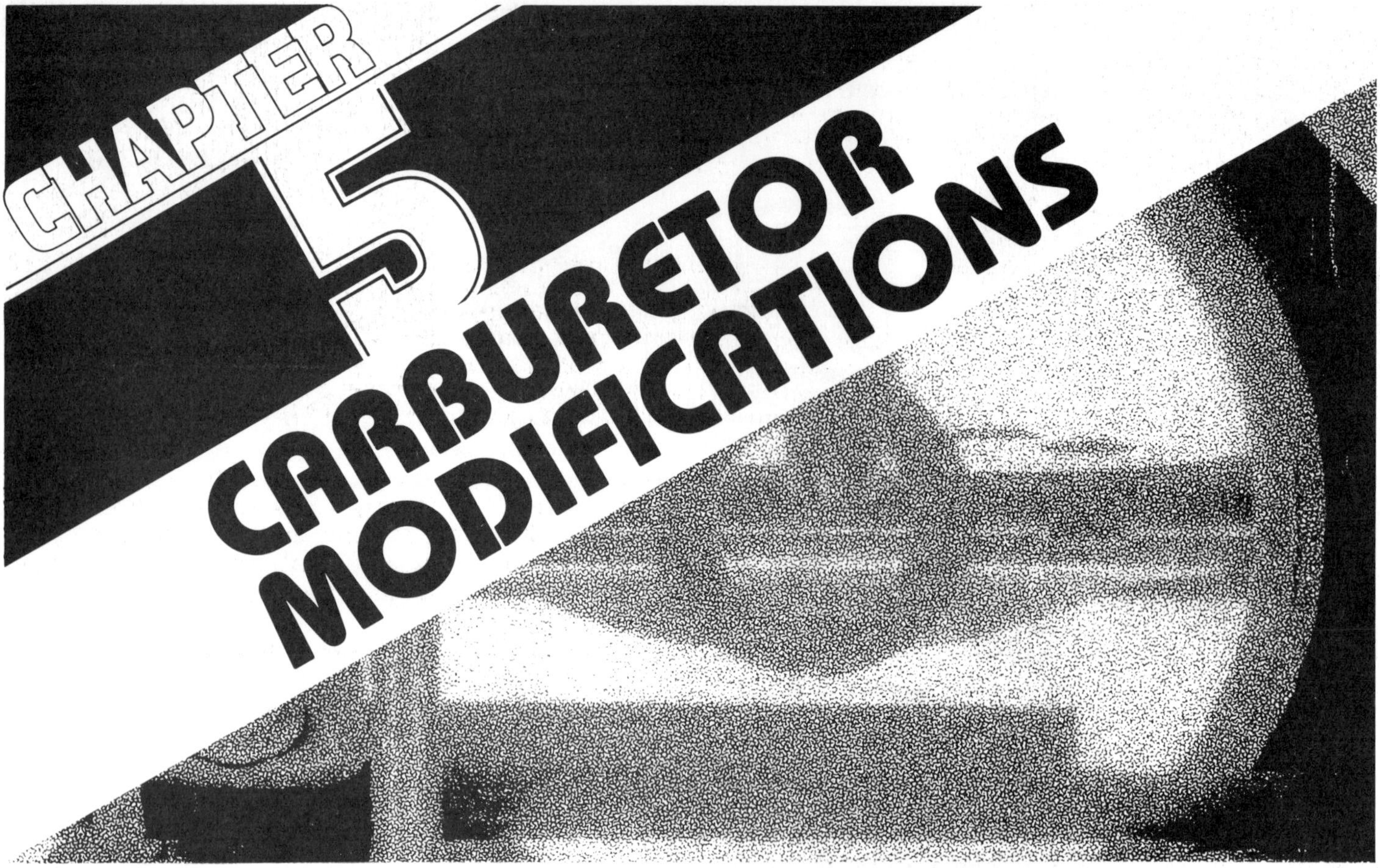

Carter carburetors "need" few modifications to provide consistent performance. Well established in the Chrysler ranks, but widely used on almost all high-performance engines, the simple, straightforward Carter design permits easy tuning, modification and repair. Of all Carter models, the Thermo-Quad is perhaps the best all around racing choice.

WHY MODIFY?

There appears to be a great disparity in the type of carburetor modifications performed by racers and performance enthusiasts; there is also an equal disparity in the degree of success achieved through those modifications. It would seem that Murphy's Law (anything that can go wrong, will go wrong) is invoked whenever a carburetor is disassembled. And more often than not, simple misunderstanding is the culprit. The "whys" of a modification must be comprehended along with the "hows," if the thrill of victory, rather than the agony of defeat, is the aim.

For openers, *the original carburetor application* dictates that calibrations fall within certain predetermined parameters. And since each carburetor circuit interacts with others, a simple modification can cause untoward ramifications, with the end result being a carburetor that suffers from "terminal dysfunction." When modifications are attempted, in the name of improved performance, driveability, or greater fuel economy, it is essential to understand Carter design philosophies if success is to be assured. And, it should be stated at the outset that *radical alterations should never be necessary for any highly specialized racing applications. The need for extensive rework is usually indicative of a damaged carburetor, or misapplication.*

Tuning and modification should be done from a "known" starting point. It's wise to check original specs before getting started. This can be done by locating the model number. Some carbs have the number on identifying tags...

Others have numbers stamped in a boss on the base plate. But wherever you locate it, jot it down for future reference; it's nice to know that you can "undo" modifications that don't yield the desired result by ordering stock replacement parts.

PRIMARY IDLE SYSTEM

Virtually all Carter carburetors, from the four-barrel Thermo-Quad to the BBS single-throat, meter idle fuel through the same type of circuit. As described in the previous chapter, *most Carters employ a standard "clockwise-lean" adjustment with needles that protrude into the curb idle discharge port.* However, some AFB and AVS carbs were produced with a single mixture adjustment screw controlling the flow through both primary barrels. Under this arrangement, bleed air, rather than air/fuel emulsion is adjusted, but mixture is varied according to the same procedure used with conventional idle systems. When a single adjustment screw is used, it is left-hand threaded so that rotating it clockwise leans the mixture. This occurs because clockwise rotation of a left-hand threaded screw causes it to back out, or loosen. Carter carburetors fitted with air bleed idle mixture adjustment also have fuel emulsion mixture needles, but these are sealed at the factory to limit the amount of fuel admitted during idle. Inspiration for this arrangement was provided by the Federal Government through its exhaust emissions regulations. Late model original equipment carburetors are restricted even further, as all adjustments are sealed.

MODIFICATIONS—WHEN AND WHERE?

In general, there should be no need for modification of the idle system other than adjustment of mixture and engine speed. However, if after a full range of adjustments are made, idle mixture is too rich or too lean, either the idle air bleed or the idle feed restriction will have to be modified. (This isn't meant to imply that these modifications must be attempted, but that they are the only options available other than replacing the carb.) Since the idle feed restrictions (also called low-speed or idle jets) on Carter carburetors are usually placed in the main well, they are all but inaccessible and all idle calibrations must be made by altering the complementary air bleed diameters. Determination of air

A bargain or big trouble? With an unknown history, carbs found at swap meets can be nothing but trouble. Since many dual four-barrel installations were used on race cars, carbs may have internal modifications. Check carefully before parting with any cash.

When adjusting the idle air/fuel mixture on emissions type carbs, it may be necessary to remove the plastic limiting caps to obtain the optimum setting. If acceptable idle quality can't be achieved within a few turns, other problems usually exist.

In addition to the throttle plates, the lower portion of a carburetor contains numerous air passages. Idle problems can result if these passages become obstructed or the base gasket used with the carb allows air leaks. Keep everything clean and use Carter or equivalent gaskets.

Selecting the correct carburetor often eliminates the need for subsequent modifications. In addition to air flow capacity, linkage and emissions connections should be considered. This 9639 AFB was developed specifically for Chrysler engine requirements.

bleed size is based on the diameter of the curb idle port, desired throttle plate position at idle, and transfer circuitry, and since all of these factors affect system timing, alteration of bleed diameter is not straightforward and is, in general, ill-advised. It is often suggested that *if the carburetor will not idle properly with factory calibrations, it is probably defective or is being used in an improper application.*

The apparent need for idle circuit recalibration arises most frequently when a carburetor originally intended for racing usage is installed on a street driven vehicle. Race carburetors are calibrated to provide a rich idle mixture as a means of compensating for exhaust dilution of the intake charge (a result of high cam overlap). When installed on a meek, mild-mannered street engine, a carburetor with race calibrations may feed excessive amounts of fuel at low speeds. This will not only affect driveability, but will seriously degrade fuel economy, hence the need to lean the mixture. However, while enlargement of air bleed diameter will provide a leaner emulsion, this change will also impact circuit timing (the point at which idle fuel flows), and if too much of the signal is bled off, the idle circuit may stop delivering fuel before the main circuit begins. This can result in an off-idle stumble that causes the car to break into an automotive version of the fox-trot.

However, if you are prepared for the consequences and there is no alternative but to drill the bleeds larger, measure the existing diameter before you begin. If too much material is drilled away, epoxy can be used to seal the air bleed, so that it can subsequently be redrilled to the original diameter—or to the size that seemed to work so well when you decided to "give it just a little more to be sure."

ADJUSTABLE AIR BLEED IDLE SYSTEMS

As previously mentioned, some original equipment carburetors utilized an adjustable air bleed to permit idle mixture alterations. These carburetors also contained a standard air bleed, but modifying its diameter serves no purpose, since this is effectively what occurs when the mixture screw is turned. When carburetors with adjustable air bleeds suffer from idle fuel delivery problems, it is because of too little fuel, not too much. The problem is complicated by later model carburetors, where the mixture screws were sealed to prevent alteration of fuel flow and exhaust emission by "monkeying around" under the hood.

A common cause of extremely lean idle is the accumulation of deposits along internal fuel passages. This restriction can limit fuel flow to the point where correct idle speed cannot be maintained. If the deposits or dirt cannot be removed by immersing the primary castings (completely disassembled) in carb cleaner, with subesquent blasts of compressed air, chances are that adjusting the mixture screws will be of no avail as the blockage will remain, and will continue to be the limiting factor in fuel flow. Always make sure that internal passages are clear of foreign material before reassembling the carburetor; use thorough visual inspection and air-nozzle tests.

All tests and rebuilding aside, the best way to prevent a "smog-lean" idle is to install an appropriate performance model carburetor—designed for wide adjustments in idle-fuel ratios.

MAIN METERING SYSTEM

In theory, if the right carburetor is chosen for a particular application, no changes in the main metering circuit, other than jetting, should be necessary. However, in some highly specialized race environments, there may be a desire to alter the booster venturis, or venturi diameter in an effort to gain a competitive advantage. Unfortunately, there is little magic to be had for the price of a screwdriver and a drill bit. And without a fully-equipped air flow bench to determine fuel delivery curves, a main metering circuit modification is usually tantamount

On WCFB carburetors, the metering rods and mechanical actuating hardware reside in a covered well in the air horn assembly. Later design AFB, AVS carburetors do not use mechanical assisted metering rods, rather metering rod actuation is entirely vacuum controlled. Some Thermo-Quad carburetors use cam assisted metering rods that ensure complete enrichment at full throttle.

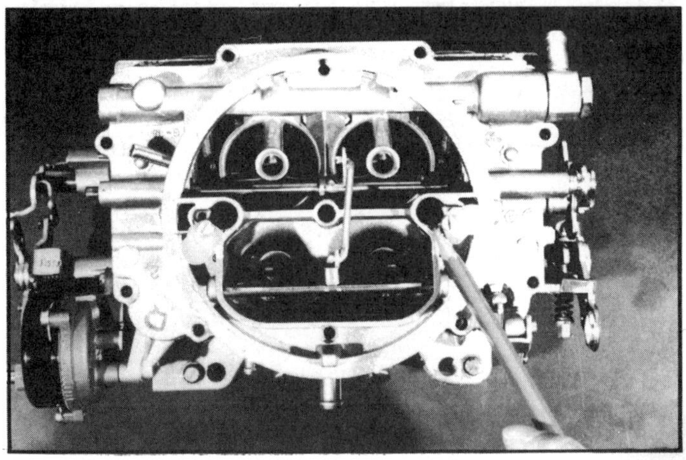

AFB and AVS metering rods are controlled by pistons that reside in bores in the airhorn assembly under small cover plates. The pistons move in the bores against springs and are directed by manifold vacuum; more throttle produces less vacuum and causes the pistons to move up in the bores, moving the metering rods out of the jets and enrichening the mixture.

to a shot in the dark. This isn't to imply that improvements are impossible, but a systematic approach backed by accurate record keeping is essential. The real problem with main metering system modifications is that they are not easily corrected, so if a mistake is made, the result will very probably be the creation of a carburetor that will function better as a paperweight. To put things in perspective, consider the following table:

MODIFICATION	EFFECT
Jet Change	Increase size = richer mixture Decrease size = leaner mixture
Air Bleed Diameter	Enlarging the main air bleed leans the mixture and delays activation of fuel discharge at main nozzle. Enlarging the idle air bleed tends to lean the idle mixture and decrease the activation signal, sometimes causing premature cessation of the idle circuit as the main system takes over. Decreasing bleed size has an effect opposite to those described above.
Venturi Diameter	Increasing venturi area reduces restriction and can increase maximum air flow but it often reduces the fuel metering signal strength.

MAJOR MODIFICATIONS

Depending upon the particular model, it may or may not be practical to perform any of the above modifications. The important and simple fact to remember is that Carter carburetors do not lend themselves well to major alterations. The AFB and AVS have removable booster venturis, but replacements are not sold separately, so the only option available is to modify existing components. Removal of material from the inside and outside diameters of the boosters will allow only a slight increase in air flow and is usually not worth the effort. However, if the need for greater air flow is vital, the venturi itself may be ground and reshaped. This may only be done on AFB and AVS models, and only on the primary side of the latter since the secondaries of the AVS are straight bores with no venturi shape (the secondary bores can be enlarged and a new air-valve fabricated for the enlarged shape, but the project requires expert-machinist talents and usually results in only minimal improvements in air flow). On the Thermo-Quad, the booster is an integral part of the air horn

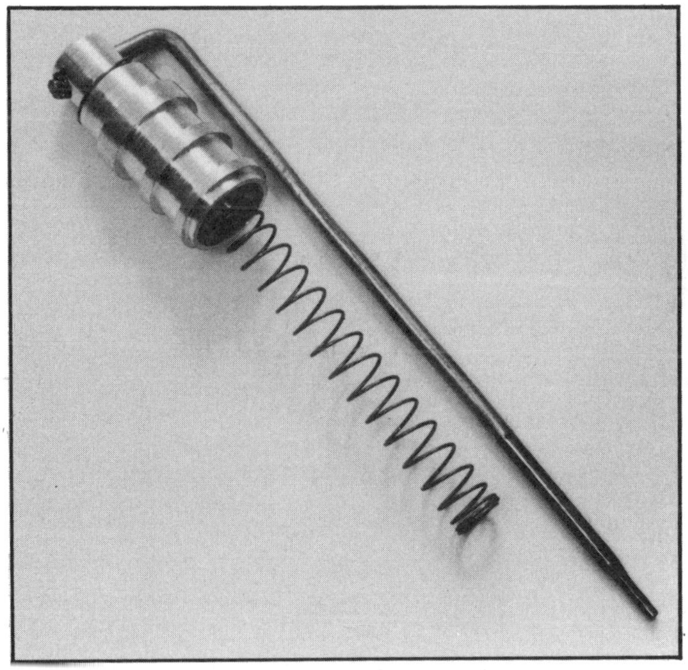

The spring protruding from the bottom of the power piston controls power enrichment timing. Carter offers several springs for the AFB and AVS.

A dimpled cover (left photo) indicates that three-step metering rods are installed. These rods also require special "high-type" jets since they must travel further as they pass through the additional metering step.

On AFB and AVS carbs, a flat cover plate (right photo) denotes the use of standard two-step metering rods. The two step rods are shorter and do not require the additional travel of the three-step variety.

casting, and when it is removed, the venturi (in the phenolic main body) is easily accessible. However, due to the nature of the material (it cracks easily), and the lack of room around the venturi, you're advised to leave this area alone.

JETS AND METERING RODS

While the foregoing may seem out of context in a chapter devoted to modifications, it is intended as an accurate appraisal of the gains to be had and the problems to be encountered when ambitious alterations are attempted. As opposed to some carburetors that are modular in construction, Carters generally utilize only two or three major castings, and consequently do not lend themselves well to major modifications. Jet and metering rod changes are the most common (and the most successful) form of main-metering system modification. Carter offers a number of convenient Strip Kits for the AFB, AVS and Thermo-Quad models, and having the appropriate part number on hand will provide a wide range of potential adjustments. These Strip Kits contain a selection of metering rods, jets and inlet needle/seat assemblies and are packaged in an easy to carry and store see-through plastic box.

If a Strip Kit is unavailable, it is a good practice to purchase an assortment of jets and/or metering rods rather than attempting to modify existing parts. Modified parts offer unpredictable performance results that are almost impossible to duplicate (and duplication is required if you desire to match the right and left side air/fuel ratios). Jets are best left undrilled as a rough finish on the orifice, or a reduction of the entry or exit chamfers can sometimes reduce—rather than increase—flow capacity, much to the confusion of the tuner.

SPARE THE ROD, SPOIL THE CARB?

In a sense, this is a very real possibility. The problem with rejetting any carburetor equipped with metering rods is that both cruise and power calibrations are affected. (This is not the case with designs that utilize a power valve and separate power enrichment circuit, as all fuel, other than what is metered by the idle circuit, passes through the main jets.) Therefore, if the jets are enlarged simply to provide a richer mixture for heavy load/full throttle use, the not-so-desirable enrichment of cruise air/fuel ratio will also result. In Carter carbs, however, this pertains only to the primary circuit, since metering rods are not used in the secondaries. In any event, the secondaries are closed under cruise conditions, so they have little—if any—effect on fuel economy.

POWER ENRICHMENT

Since it is the area, rather than the diameter of the jet, and the restriction (i.e., the metering rod) that must be considered when computing the effect that a change makes on air/fuel ratio, a series of mathematical computations must be completed in order to accurately determine the effects of a jet

Staged power enrichment, as offered by three-step metering rods, is especially helpful when an engine is used to propel a heavy vehicle such as a motorhome. Partial enrichment at moderate loads minimizes lean-misfire problems. In addition, the phenolic body of the Thermo-Quad is a real asset in hot engine compartment applications; and it doesn't get much hotter than in a van/motorhome.

change on both power and cruise mixtures. One must first compute the area of the existing jet and the area of both steps of the metering rod, and then subtract the latter from the former to determine the *effective jet area* at each step in the metering rod movement. After the jet is changed, the new jet area must be calculated, and the rod areas subtracted to yield the *new effective areas*. The difference between the old and new effective areas would then determine the percentage change in air/fuel ratio.

For example, consider a number 9625 AFB which is equipped with .098-inch diameter primary jets, and a metering rod with an economy step of .0745-inch, and a power step of .047-inch. Using the standard formula for area:

Jet Area = PI X $(\text{Jet Diameter}/2)^2$
Jet Area = 3.1416 X $(.098/2)^2$
Jet Area = 3.1416 X $(.490)^2$
Jet Area = 3.1416 X 0.2401
Jet Area = 0.00754-square inch

Applying the same formula to the metering rod diameters of .0745- and .0470-inch:

Metering Rod Area = 3.1416 X $(.0745/2)^2$
Metering Rod Area = 0.00436-square inch
and
Metering Rod Area = 3.1416 X $(.0470/2)^2$
Metering Rod Area = 0.00174-square inch

Now, subtracting the metering rod areas from the jet area (.00436 and .00174 from .00754) yields effective areas of .00318- and .00580-square inch for economy and power steps respectively.

With the equivalent areas of the power and economy steps known, we have a baseline from which changes can be made. For example, if it was determined that the power mixture should be enriched by 5% but the economy step should remain unchanged, then a metering rod with a smaller power step (smaller than .047-inch) and the same size economy step (.0745-inch) should be chosen. But Carter does not offer an unlimited choice in metering rods, and a rod with a smaller power step than .047-inch is not available. There are two ways to overcome this obstacle: (1) a larger jet and metering rod will have to be installed that reestablishes the original rod-to-jet area relationship for economy operation (a larger rod and jet diameter on the economy step yielding the same .00318-square inch effective area), but the new jet and rod must produce a 5% larger area on the power step. In this case a 5% richer jetting would require a power step area of 1.05 X .00580-square inch or .00609-square inch); or (2) you can machine the power-step end of the metering rod. This can be done by either twisting the rod between two surfaces of fine emery paper, or mounting the rod in a drill press while the end is held against emery paper or a flat file. If you choose to modify the metering rod, a precision micrometer should be used to determine that both metering rods have been modified correctly (both rods have the same size and are correctly tapered over the length of the power step).

The theory of calculating the new metering rod diameter or jet size from the desired step area can be explained most clearly by following through an example of situation (2) from the above paragraph. In this example a 5% richer mixture is desired. (This equates to a 5% increase in the fuel-flow area when the metering rod is raised for power enrichment). This 5% increase in area is calculated by multiplying the original area by 1.05 (a 10% increase would require a factor of 1.10); in this case .00580-square inch X 1.05 = .00609- square inch—an enlargement of .00029-square inch (.00609 -.00580). Therefore if the metering rod area is *reduced* by .00029-square inch on the power step, the fuel-flow area at the power step will increase by 5%, producing the desired richer mixture. Since the original metering rod power step area is .00174-square inch and a reduction of .00029-square inch is required, the new step area must be .00145-square inch.

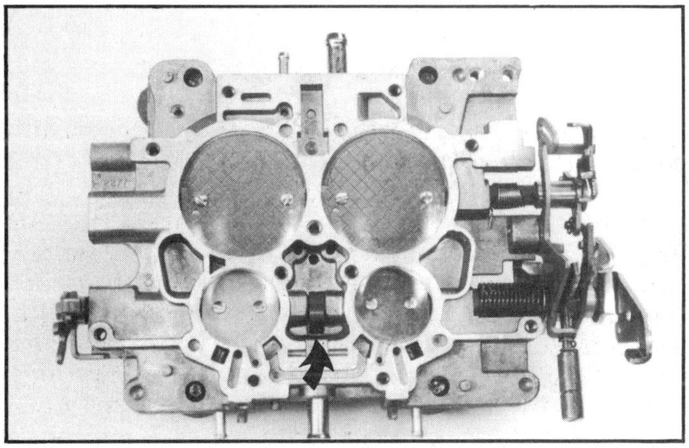

The cam-shaped piece situated between the primary throttle bores (arrow) provides mechanical assistance in actuating the power enrichment circuit when the carb is brought to the wide-open-throttle position. This device is used only on Thermo-Quads.

Power enrichment timing is controlled by small springs that operate against the metering rod pistons. Alterations in timing require replacement or modification of these springs. Trial-and-error experimentation is required to determine the optimum setting.

Now that the new metering rod area is known, the diameter that will equal that area must be calculated. This can be done by modifying the original area equation:

$$\text{Rod Diameter} = \sqrt{(4 \times \text{New Area})/3.1416}$$

Using this equation, we can convert the metering-rod area to an equivalent diameter. Plugging in the numbers:

$$\text{Rod Diameter} = \sqrt{(4 \times .00145)/3.1416}$$

$$\text{Rod Diameter} = \sqrt{.00580/3.1416}$$

$$\text{Rod Diameter} = \sqrt{.00185}$$

$$\text{Rod Diameter} = 0.043\text{-inch}$$

So a rod size of .043-inch—or a .004-inch reduction in the power step—will produce a 5% increase in fuel flow.

A MATTER OF TIMING

In Carter carburetors power enrichment is provided by a piston/spring assembly that raises and lowers the metering rod positions in the primary jets, upon reaction to engine vacuum. For AFB and AVS carburetors, Carter offers a piston spring assortment under part number 61P-1083U. The load ratings of these springs are 6, 7-1/2 and 9 ounces, which tells little about the effect on enrichment timing except that the weaker springs will require less force to overcome their pressure, hence they will lift the metering rods (enrichening the air/fuel ratio) at a lower manifold vacuum reading. Looking at it with reverse logic and using the 6-ounce spring as an example, only 5 or 6 inches of vacuum (rather than 7-10 for the other springs) will compress the weaker spring, thereby "closing" the power-enrichment circuit (leaning out the air/fuel ratio for economy operation).

Since there is little reliable data on the correlation between spring load and engine vacuum, the trial-and-error testing method must be invoked. In addition to changing springs, it is also possible to alter pressure by adding a shim, which effectively stiffens the spring, or by removing a few coils, which has the opposite effect. Since Carter doesn't offer a spring kit for the Thermo-Quad, this is the only available approach. However, it should be noted that the TQ has a three-step metering rod, so the effect of spring changes on the intermediate step mixture (part-throttle load) must also be considered.

The concept of tailoring the metering rod timing is rather straightforward, but it is truly amazing that both backyard tuners and allegedly competent race car mechanics set about changing power enrichment timing points without the faintest notion of the manifold vacuum for the engine under test. The reason for such misunderstanding often arises from the simple lack of a vacuum gauge. Such a device is the only means of peering into the internal operating environment of the engine. A reliable

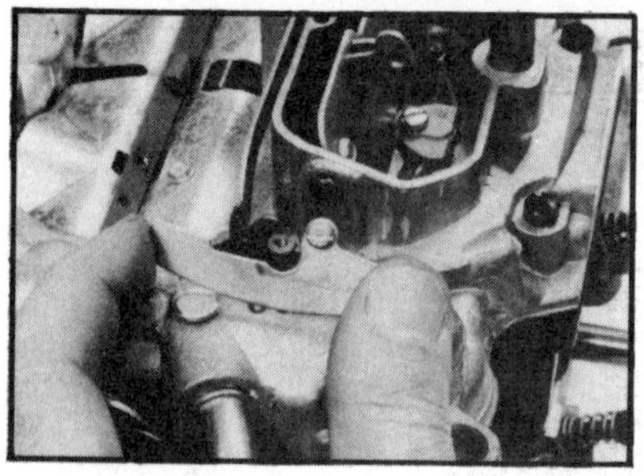

This plate guides and protects the metering rods on Thermo-Quad carburetors. It should always be replaced before the returning the carb to service; even during those "between-run thrashes."

vacuum gauge is absolutely essential when dealing with a metering rod equipped carburetor. *Alteration of power circuit timing is a trial-and-error effort and reliable reference points can only be determined with a vacuum gauge.*

DISCRETION—THE BETTER PART OF VALOR

Selection of power circuit opening point should be based on the amount of manifold vacuum present during "normal" operation. Obviously "normal" covers quite a broad area, depending upon engine application. Where a large-displacement engine in a lightweight street vehicle may cruise at a vacuum reading of 18-in/Hg, and rarely pull the needle below 6-in/Hg (except during full-throttle acceleration), an RV or motorhome will be lucky to cruise at 9-in/Hg and will pull 18-in/Hg only when rolling down a long grade loaded with kids, inlaws, dogs, toys and a port-a-potty. With such a wide range of potential operating conditions, there is no "right" or "wrong" setting for the power-circuit opening point, it simply must be correlated to engine requirements.

Unfortunately (or fortunately) the bottom line is: in terms or "major" performance gains, fussing with metering rod timing may be more trouble than it's worth. However, in cases where an engine is mounted in a heavy chassis, or in a vehicle used for towing, bringing the power circuit into play at a higher manifold vacuum reading can reduce or eliminate part-throttle knock. Quite a different situation exists with a high-performance or race

When dual four-barrels are used, carb tuning becomes more difficult because manifold vacuum at wide open is extremely low; this is particularly true of jetting for uniform fuel distribution and metering rod piston spring modifications for enrichment timing. Be careful to maintain uniform power enrichment in both carbs.

Proper power enrichment circuit timing can be very elusive when exotic intake systems are used. This twin turbo installation produces a strong vacuum signal in the carburetor plenum and can maintain a "light-throttle-lean" condition during fairly hard acceleration without modification to the metering rod piston springs.

The real benefits of proper power enrichment timing (and three-step metering rods) are to be found in heavy vans, motorhomes, and vehicles that tow trailers. In these applications carefully tailoring the air/fuel ratio to manifold vacuum can maximize fuel economy and throttle response, while minimizing ping and overheating.

engine equipped with a long-duration cam. Manifold vacuum at idle may be so low that the power enrichment circuit is always open (i.e., the metering rods are never pulled into the economy position). In this situation, reducing the metering-rod spring pressure will allow the rods to return to the economy position. The low speed circuit will consequently provide a substantially leaner mixture and probably greatly improved part-throttle engine response and gas mileage. In these cases, alteration of metering-rod timing can have a significant effect, but otherwise it's pretty much a matter of fine tuning, and the effort required may not be justified by the benefits received. It all depends upon how far the standard setting misses the mark.

In theory, the metering rods should rise to the power step when the manifold vacuum is just below the lowest reading encountered within the normal-cruise/light-acceleration range. Viewing the power circuit as an "extra gas switch," the switch should be turned on whenever the engine will be placed in a demand-for-power situation. If the opening point is too high, extra fuel will be added during the momentary vacuum drops that occur during light acceleration, needlessly reducing economy.

However this "ideal" is not the whole story. The power enrichment circuit is sometimes opened early, i.e., at a higher vacuum reading than is found during high-load situations, to get an additional supply of gasoline moving. This "tuning" is performed more in response to an anticipated—rather than current—engine demand. Consider the characteristics of a typical V-8 engine. At idle, with no load applied, manifold vacuum will read in the 16- to 18-in/Hg range. During normal driving conditions, the reading will vary from approximately 9 in/Hg under light acceleration to 17 in/Hg at cruise. Only during faster-than-normal acceleration, as is encountered in speed contests, taxi cab rides, or driver "over-enthusiasm," will the vacuum drop below 9 in/Hg. Therefore, the typical street engine will spend most of its life in the 9- to 17-in/Hg range. Consider these parameters:

1) A 10- to 11-inches of vacuum "opening" point would richen the mixture unnecessarily during normal acceleration, causing a deterioration in fuel economy.

2) Timing the metering rods to raise at 6 to 7 in/Hg would leave a fuel delivery "hole" with no mixture enrichment between the high-load point at 6 in/Hg and the transition to the above 9-in/Hg cruise range. This delay in power enrichment can create stumble, misfire, and ping.

3) Considering the above, activation of the power circuit at 9 in/Hg would be nearly ideal since it

Even with two large AFBs sitting atop an engine, reasonable fuel economy can be obtained with proper jetting and calibration of enrichment timing (and, of course, using a light throttle foot).

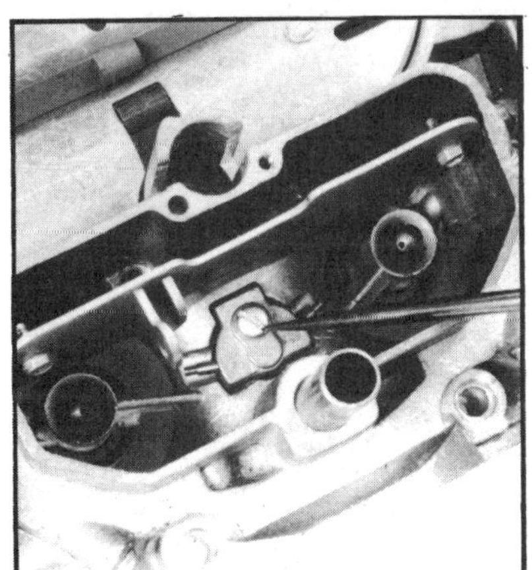

The accelerator pump nozzle is often overlooked as the cause of an off-idle stumble. The nozzle diameter affects the duration of pump shot (how long to pump the same volume of fuel). Enlarging the nozzle will squirt the pump volume faster. The *volume* of the pump shot can be altered only by changing the rod link position in the pump arm (closer to the pivot pin pumps more fuel).

initiates additional fuel flow precisely at the point where the engine is beginning to require a richer mixture.

As in most things automotive, trial-and-error experimentation will prove to be the final acid test. Unfortunately, with the continuing deterioration of gasoline quality, repetitive experimentation may be required to maintain optimum performance and economy over the long term. In many instances it is necessary to initiate power enrichment before the optimum vacuum reading in order to prevent ping under partial load. While the engine might not be laboring noticeably, the load associated with 8-10 inches of vacuum is frequently sufficient to launch a normally respectable gasoline engine into a cheap impersonation of a Mercedes diesel.

It is for precisely this reason that the three-step metering rod was developed. Extra fuel volume is admitted by the intermediate step and fills the need for a slightly-richer-than-normal mixture when the engine is under light load. With two stages of enrichment, partial and full, a lean cruise mixture can be maintained while surging, stumbling and misfire under light load, are eliminated. The three-step rod will not affect or slightly enhance gas mileage. By bringing about partial enrichment at 9-12 inches of manifold vacuum, it is often possible to reduce main jet size (as compared to the requirement for an equivalent carburetor with single-stage enrichment) while maintaining acceptable driveability.

TUNING THE ACCELERATOR PUMP

Carter uses a rod and link arrangement to actuate the accelerator pump. This system can be "tuned" to alter pump characteristics by bending the rod, or reinstalling it in a different hole in either the upper or lower actuation levers.

Discharge volume is controlled by the piston at-rest position (the position of the accelerator pump in the pump bore when the throttle is closed). As the at-rest position is raised, more fuel will flow into the pump chamber below the piston. When the throttle is fully opened, the piston will discharge the full volume of fuel within the pump chamber. However, if the piston is raised above the fuel level in the float bowl, a delay in discharge will result since an air gap is introduced between the piston and the fluid in the pump chamber. For this reason, float-level setting is important, as a low fuel level can result in stumble during acceleration, caused by a "leaner" mixture of air-diluted fuel in the pump shot.

For optimum performance and economy—with fuel level properly set—the accelerator pump piston should be positioned to discharge just enough fuel to provide smooth acceleration. An over-generous pump squirt is almost as bad as not enough, as plug fouling, poor fuel economy, and sluggish throttle response are usually the results. It should be remembered that regardless of application, the pump circuit should provide just enough fuel to prevent stumble. Once this point is reached, the delivery of additional fuel will degrade overall performance, particularly while the engine recovers from an over-rich condition.

CHANGE THE NOZZLE, ALTER THE RATE

Circuit timing, however, rather than fuel volume, is frequently the culprit that's causing a stumble. With a given volume of fuel to be discharged (that fuel contained within the pump chamber), and a piston—driven by a spring—that discharges through a fixed nozzle orifice, *it is nozzle size that affects the fuel discharge rate* (how fast the fuel is pumped into the engine). Squirting in all the gasoline in the float bowl still won't overcome a stumble if it is all consumed before the main system starts functioning. The idea here is to size the nozzle so that the accelerator pump continues to squirt fuel throughout the entire period during which the main system transitions from non- to full-function. As nozzle diameter is decreased, the discharge duration is lengthened, and vice versa. Carter offers nozzle clusters ("squirters") in a variety of sizes, and if availability is a problem, the nozzles can be drilled with few of the ill effect the accompany main jet drilling. *Nozzle size has no appreciable effect on total discharge volume, only the amount of time required to discharge that volume.*

When adjusting discharge duration, it is advisable to have several "squirters" on hand. Carter offers a nozzle package under two part numbers; 48P-379U applies to AFB's and 48P-380U applies to Thermo-Quads. Three nozzles (also called accelerator pump clusters) are contained in each kit. Each one should be tested to determine optimum performance.

SECONDARY MAIN METERING SYSTEM

Except for the Thermo-Quad, all Carter four-barrels use the same type of main jets on both primary and secondary circuits. Although there is usually a considerable size difference, owing to the use of metering rods in the primaries, the jets are interchangeable. Secondary rejetting is simply a matter of swapping the existing jets for others of different size.

Secondary jetting on the Thermo-Quad is essentially the same with the Thermo-Quad, except that primary and secondary jets are not interchangeable. The secondary "jets" are actually quite different in appearance, being "long hunks of brass," rather than the more customary short, threaded insert. The unique secondary jet configuration owes in part to the Thermo-Quad's phenolic main body, and the minimization of passages within it. Since fuel must be brought from the float bowl floor up to the discharge nozzle, a path must be provided, and Carter uses the jet to complete the circuit. Rather than screwing into the float bowl floor, the jet threads into the aluminum upper casting (the body that also houses the choke), and extends down to the bowl

With its huge secondary throttles, the Thermo-Quad can be tuned to provide excellent drag strip performance. Air flow capacities are 800 or 1000CFM.

floor. It might best be thought of as a metering orifice at the end of a straw. Aside from its unusual appearance and the fact that is is held in position with an O-ring, it is identical in function to a standard screw-in jet.

NUMBERING SYSTEMS—THERE'S LOGIC SOMEWHERE

Carter utilizes several part-numbering systems for jets, all of which contain information relating to orifice size, but they utilize a unique encoding. All WCFB's, most AFB's and AVS's and 1974 and later BBD carbs use jets with a three digit suffix that describes orifice size. (All Carter jet part numbers begin with "120-".) If the suffix begins with a "3" jet size is less than .100-inch; a "4" in the first suffix position indicates an orifice of .100-inch or greater. Thus a .098-inch jet would carry part number 120-398 while a .101-inch jet would be labelled 120-401. If you're still with us, now we'll get tricky. When referencing the jets used in AVS and AFB carburetors with three-step metering rods, a "5" in the first suffix position pertains to sizes over .100-inch and a "4" describes jets with diameters of .099-inch or less.

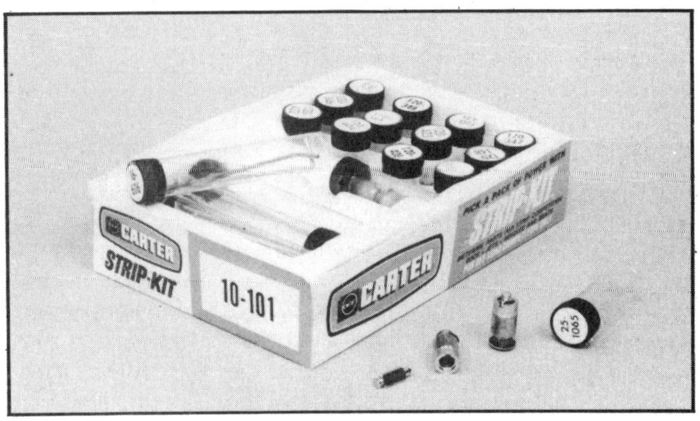

Carter's Strip Kits contain an assortment of jets, metering rods and needle-and-seat assemblies. They take some of the difficulty out of carb tuning chores.

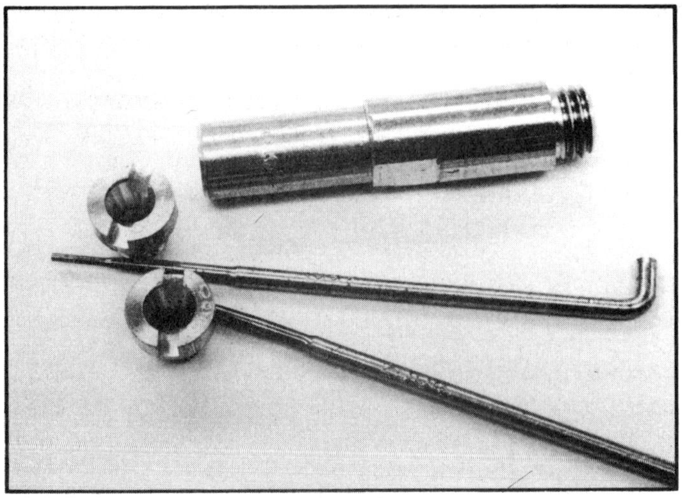

Carter's numbering system is somewhat tricky. The numbers relate to jet size but are "encoded;" without knowing the key (see text) a potential carb tuner is in for a real head-scratching session.

Thermo-Quad jets have a numbering system all their own, utilizing a four digit suffix with either a "4" or "5" in the first position. The "4" relates to primary jets while a "5" indicates a secondary jet. Diameter is encoded in the next three digits thus, part number 120-4107 is a .107-inch primary jet. This numbering system applies to all Thermo-Quads except the 4000 series (which includes the Competition Series). These carburetors have two, rather than three step metering rods, and a jet part numbering scheme that defies comprehension. If you intend working with this carburetor, only the metering rods should be changed. Part of the problem comes from the fact that two of the Competition Series Thermo-Quads listed in the (Strip Kit cross reference section) Carter catalog are equipped with pressed-in rather than screw-in jets. These are marked with a three digit suffix, making the part numbers 120-331 and 120-334. With pressed-in jets, all calibration changes must be made through a swap of metering rods.

Although secondary air valves are used in all Carter four-barrels, the throttle plates themselves are opened mechanically. Ensure that both the primaries and secondaries are in the fully-open position when the throttle linkage is against its stop.

To adjust wide-open throttle-plate position, bend the rod that connects the primary and secondary linkages. The arrow indicates the rod to adjust on a Thermo-Quad.

SECONDARY ACTUATION

Except for a few limited production versions of certain models, all Carter four-barrel carburetors utilize either a counterweighted auxiliary throttle valve or a spring-loaded air valve to control air flow through the secondary bores. Conceptually, both arrangements are identical, and function in conjunction with a mechanically actuated secondary throttle to provide smooth, stumble-free transitions to wide open throttle. Since the auxiliary throttle valves used in WCFB's and AFB's do not completely seal against the bore, there is always some air flow initiated just as soon as the mechanically operated throttle valves are opened, and as air flow increases, the counterweights are overcome and flow increases further.

With the auxiliary throttle valve counterweights controlling the actuation of the secondaries, in order to speed the secondary opening, some material must be removed from the weights. This is a simple modification as the weight is externally accessible on WCFB's, and may be easily removed from AFB's. Only small amounts of weight should be removed, as a counterweight that is too light will allow the air valve to flop open prematurely, causing the engine to stumble or hesitate.

In most cases, the counterweight is properly calibrated so that no modification is required. However, the larger the engine displacement in relation to the carburetor air flow, the more likely that throttle response will benefit from lightening the counterweights. And in some instances, especially where a low air flow carburetor is used on a large engine, it is possible to entirely remove the auxiliary throttle valve. However, before depositing it in the trash can, the car should be driven to ensure that the carburetor may be smoothly brought to wide open throttle, with the transmission in high gear (a condition of maximum load, most likely to cause a stumble).

AIR VALVE ADJUSTMENT

Adjustment of the secondary air valve used in the AVS and Thermo-Quad carburetors is much easier than modifying the counterweights used on auxiliary air valves. Since spring pressure holds the air valve closed, altering that pressure increases or decreases the opening rate. Both the AVS and the Thermo-Quad carburetors have a locking screw which must be loosened before the spring-adjusting screw can be turned. Whenever making an adjustment, the spring screw should be held while the lock screw is being loosened to prevent the spring from unwinding. If unwinding occurs, a new reference must be established and subsequent adjustments made from that point.

Bearing in mind that carburetor "tuners are not without "human error," developing a repeatable reference point *before* a trial-and-error tuning session is begun can save many hours of lost effort" caused by an unwound air valve spring. Many racers form an initial reference point by unwinding the spring until the air valve rotates into a vertical position, then tighten the spring a particular number of "turns" from that point. In this case if the spring accidentally unwinds, it can be returned to approximately the same position.

Two screwdrivers are the only tools required to adjust an AVS air valve spring, and while the same arrangement will work on a Thermo-Quad, life is much easier when Carter's special tool, part number 109P-397, is used. The Thermo-Quad spring adjustment screw is located within, rather than next to the lock screw, and although a narrow-bladed

Adjusting the spring tension of Thermo-Quad air valves is most easily accomplished when this tool is used in conjunction with a garden variety screwdriver. This Carter tool allows spring tension to be maintained while the lock screw is loosened. Adjustment on the AVS can be accomplished with just a screwdriver.

One of the Thermo-Quad metering rod cover plates contains a tab that serves as an air valve stop. If it is removed, the valve can go over-center, reduce air flow, and disturb fuel distribution.

screwdriver can be used to hold and/or turn the lock screw (once it has been broken loose with a wide-bladed screwdriver) while another screwdriver is used to hold the spring screw, three less hands are required when the Carter tool is used.

For applications where performance is of primary interest, spring tension should be adjusted to provide the quickest possible stumble-free opening rate. A considerably greater amount of tension will be required to provide smooth transistion from the primaries to the secondaries if the vehicle is a heavyweight, since the engine loads encountered may be sufficient to require full opening of the primary bores—and therefore partial opening of the secondaries—when climbing a hill or accelerating a bit more rapidly than normal. The resulting air flow in the secondaries may open the air valve when not desired (and produce the well-known "bog") if spring tension is too light.

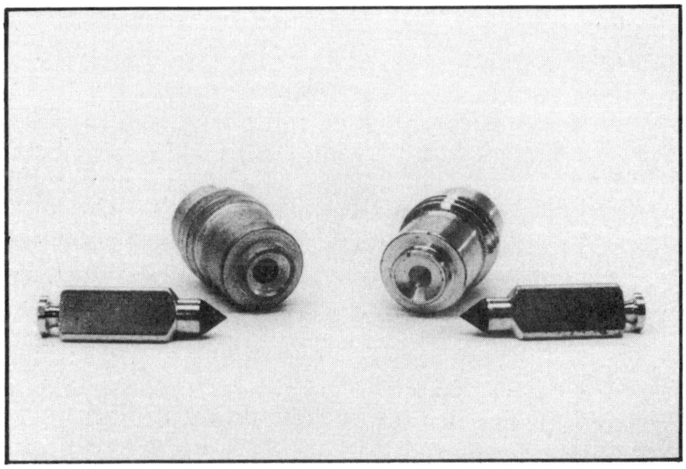

Here is a commonly overlooked source of carburetor problems. An undersized inlet needle-and-seat assembly can lead to fuel starvation problems at high engine speeds, particularly on large-displacement engines.

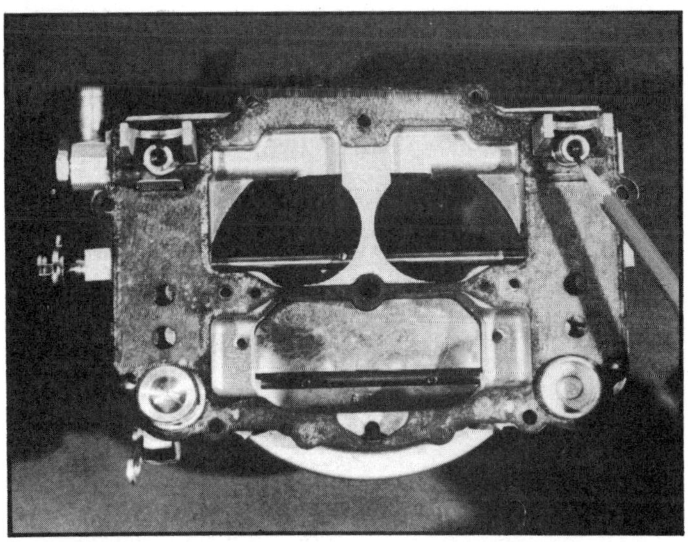

The seats are then simply unscrewed from the carburetor top. The seats are made of brass, so use a wide-bladed screwdriver to avoid damaging the screw slots.

FUEL INLET SYSTEM

Without an adequate fuel supply, the best carburetor in the world will be reduced to a metering nightmare. It's much like an athlete who doesn't eat the proper food in adequate amounts—he simply can't perform at 100% efficiency. Likewise, carburetor fuel metering circuits operate at less than optimal levels when there is insufficient fuel to go around. As a solution to this problem, many well-meaning carb tuners install the largest needle-and-seat assembly available, eliminate every filter and baffle in the system, and then proceed to scratch their collective heads when the situation progresses from bad to worse.

In general, very little in the way of modification need be done to the fuel inlet system of any Carter carburetor. As a minimum, larger needle and seat assemblies may be installed to insure that there is no

Changing to a larger inlet valve involves requires removal of the carburetor top and float hinge pins (arrows). Once the pins are out, the floats and needles can be removed.

Whenever a needle-and-seat or float assembly are removed or changed, float level and drop should be checked and reset if necessary. This level checking gauge was fabricated from a piece of thin cardboard and was supplied in a carburetor rebuilding kit.

bottle neck as fuel enters the carburetor. Most AFB, AVS and Thermo-Quad models are fitted with .0935-inch or .1015-inch assemblies at the factory. These sizes are far from restrictive for most applications, but larger diameter pieces are available. However, with this size of inlet seat, insufficient fuel delivery is almost always the result of problems located elsewhere in the system. With adequate fuel pressure, only a float level that is set too low, clogged inlet filters, or dirt lodged in the inlet seat can cause the carburetor to "starve." More commonly, the trouble lies in a crimped line, sediment infested gas tank, or an inadequate fuel pump.

For typical high-performance use, installation of the optional .111-inch needle-and-seat assembly and setting float levels as recommended should provide trouble-free operation. When either the AFB or Thermo-Quad are used for race applications, a larger .120-inch or .125-inch inlet valve assembly may prove advantageous. These will provide for fuel flow of up to 110 gallons per hour (assuming there is sufficiently capable fuel pump to back them up).

In addition to float level, float drop must also be checked. Generally, a small tang on the float determines the maximum distance that the float may drop (lower into the float bowl) and this determines the distance that the needle is pulled away from the seat, affecting fuel flow through the needle and seat assembly. If this tang were maladjusted, it would be analogous to an intake valve being only partially opened; flow is understandably reduced. Float drop may be easily checked by simply holding the air horn assembly and first visually inspecting needle position with the float tang resting against the inlet seat. In this configuration, "float drop" is measured distance from the float to the air horn assembly. In most situations, too much float drop rarely causes problems; To adjust, simply bend the tang to allow the float to drop to a lesser or greater degree, as required.

RACING MODIFICATIONS

It is a basic fact that Carter carburetors are not easily modified for all-out competition. As a result,

Carter carburetors are sometimes overlooked by racers. However, the Carter design is simple, reliable, and effective, and with a little expert help—being administered here during a dyno test in the Jenkins Competition shop—the Carter will perform alongside any other carburetor in most applications.

few companies offer reworked Carter four-barrels or modification services. This situation might lead one to believe that nothing can be done to coerce a race engine equipped with an AFB, AVS or Thermo-Quad into producing more power. Fortunately, that's not true. Race car performance *can* be improved by a few carburetor alterations that will either increase air flow or improve fuel-handling capabilities.

Of all Carter four-barrels, the AFB lends itself most readily to major modifications. The AVS is a close second, but the Thermo-Quad is a distant third as its phenolic mainbody and double primary booster venturi make machining difficult or impractical.

A LITTLE HERE, A LITTLE THERE

A slight air-flow improvement may be achieved by altering the booster venturis, which are fitted to the primary throttle bores of AVS and all throttle bores of AFB carbs. In addition to "cleaning up" the outside diameter of the booster itself, it is also possible to narrow its support leg, which protrudes into the air

One of the best "racing modifications" is installing one of these—a three-barrel AFB. It was developed especially for Pontiac and requires use of a special manifold to accommodate the secondary throttle plate.

A slight increase in air flow capacity can be achieved by machining the booster venturi. Modified version is at left. Note the leg has been both narrowed and shortened and the booster has been smoothed.

This view shows the top side of the secondary boosters in an AFB. The one on the left is standard, while the one on the right has been machined. Note the extent to which the leg has been narrowed.

stream. This work is best done with a small high-speed grinder. These tools will remove material quickly yet may be easily controlled to prevent cutting too deeply. A word of caution: the boosters are aluminum castings which have brass tubes running through their centers. When the main metering circuit is activated, the air/fuel emulsion flows through these tubes, the ends of which serve as discharge nozzles. If too much aluminum is removed from the booster support leg, it is possible to grind right into and through the brass tube, thereby turning a perfectly good booster into a useless fishing weight.

Narrowing the boosters isn't the only means by which to increase air flow—reducing the *height* of the support leg will also help as it reduces the friction to be overcome by the air stream. Again caution is in order as grinding to far will damage the brass fuel tube. While working on the leg, be careful not to grind the bottom of the booster venturi itself as the fuel metering signal may be adversely affected. Once the leg has been properly shaped (a modified booster is illustrated in the accompanying photos), blend it into the outside diameter of the venturi with a smooth radius and carefully remove the casting lines. When all grinding has been completed, smooth the altered surfaces with fine grit sandpaper. It should be noted that all this work will result in only a minor increase in air flow so don't expect a miraculous improvement in performance.

A LITTLE MORE HERE, FOR A LITTLE MORE THERE

Air flow potential can be further enhanced by removing the choke and its attendant linkages. (Carter offers two Competition Series AFBs without chokes.) On the AFB, AVS and Thermo-Quad, choke plate position is controlled by a link rod above the secondary venturis. Therefore, when the chock is eliminated, these obstructions are removed from both the primary and secondary barrels.

Choke plates are held in their shafts by two brass screws which are staked (flattened with a punch) on the threaded end to prevent them from vibrating loose. When removing these screw, increased

Removal of the choke plate will also increase air flow, however, the improvement is generally minor, albeit easy. Aside from the missing choke plate, this looks like an ordinary AFB. Look again and you'll notice the fuel fittings on both sides of the carb, making this a dual-feed AFB.

Carter offers excellent high-capacity mechanical fuel pumps for most engines. Adequate fuel delivery is the first step to top performance and reliable tuning modifications.

The Carter electric fuel pump is rated at 72 gallons per hour and is internally regulated to provide 7 psi of fuel pressure. It is very well designed, reliable, and relatively inexpensive.

resistance will be encountered as the staked portion passes through the threads in the shaft. It isn't necessary to grind the staked threads; continued rotation in the counter-clockwise direction will usually "reform" the screw end as it is extracted from the choke shaft. But this will only work properly if the screwdriver tightly fits the screw slot. Again, the resultant airflow improvements will not be dramatic, but every little bit helps.

Don't be too hasty to deposit the choke pieces in a trash bin—you may want to reinstall them at a later date. If you do, restake the choke plate screws—after they have been cinched down—by repeening or "squashing" them with a pair of vise grips. If you use a hammer and punch, be sure to place some form of support under the choke shaft so that it isn't bent or damaged in the operation.

MORE FUEL FOR THE FIRE

A race engine demands fuel like a wino demands distilled spirits. Consequently, the fuel delivery system must be reworked to ensure that it is not restrictive. As such, a high-capacity fuel pump is a necessity. Carter offers an electric pump rated at 72 gallons per hour with internal regulation to provide a steady pressure of 7psi. The company also offers several excellent mechanical fuel pumps rated at 80 gallons per hour with pressures ranging from 6 to 8.5psi. Holley, Moroso, Mr. Gasket, and other suppliers also offer high-volume pumps suitable for competition use.

With a minimum of 3/8-inch fuel lines connecting the gas tank, fuel pump, and carburetor, the next step toward ensuring adequate fuel during "pedal to the metal excercises" is reworking the carburetor inlet circuit for maximum capacity. Most WCFB, AFB, AVS, and Thermo-Quad carbs are originally fitted with needle-and-seat assemblies having a .0935-inch diameter orifice. (Part numbers are 25-1086 for Thermo-Quads and 25-860 for the others). For competition work, Carter offers needle-and-seat assemblies with a .120-inch orifice for the WCFB, AFB, and AVS (part number 25-862), or a .125-inch opening for Thermo-Quads (part number 25P-1091). However, the biggest isn't necessarily the best since the larger needle may be difficult to seal with high inlet pressures—often requiring readjustment of the float level—as a higher fuel level may develop in the bowl before the needle is forced against its seat with sufficient pressure to stop fuel flow. So some experimentation with intermediate sized inlet seats is suggested.

DUAL FEED

The fuel demands of a fire-breathing race engine may be better satisfied by a dual-feed fuel inlet conversion; a modification readily performed to AFB and AVS carburetors. The air horn casting of these carburetors is designed to allow fuel to be admitted from either the left or right side. Carter uses only a single fitting, so bringing dual-feed happiness to an AFB or AVS is simply a matter of drilling and tapping the unmachined inlet boss. The fuel passage itself measures approximately 3/8-inch in diameter. Once this hole is drilled through, the inside diameter of the boss may be threaded with a 9/16-18 tap. This will allow installation of either a standard Carter fuel line adapter (part number 156-198) or the appropriate AN fitting if braided line is used. This modification should be performed by a competent machinist with the appropriate equipment (i.e., mill, drill press, etc.).

The best fuel pump in the world won't be able to push an adequate supply of fuel to the carburetor if the lines are too small. Braided stainless steel hose is available in many diameters and is also quite attractive. This beautifully done Paxton installation illustrates the "right" way.

Trial and error is the only effective way to fine tune a carburetion system. Accelerator pump volume, jet size, metering rod selection, and even fuel pump pressure affect throttle response and horsepower output. The safest and most accurate (runs are timed) testing is done at the race track.

FINE TUNING

Developing "optimum performance" is possible only through a series of trial-and-error tests. Carburetor tuning is greatly facilitated by the "Strip Kits" Carter offers for the AFB, AVS and Thermo-Quad. In conjunction with the metering rods, jets, and needle-and-seat assemblies contained in these kits, an additional selection of accelerator pump discharge nozzles and metering-rod piston springs should be obtained.

Drag strip tuning will probably require the installation of an accelerator-pump squirter nozzle with larger-than-stock orifices. Carter offers nozzles (also called clusters) in three different sizes. However, three sizes may not suit all requirements, so it may be necessary to drill the orifices to obtain the correct flow requirements.

Alteration of orifice size does not affect the *volume* of fuel discharged through the accelerator

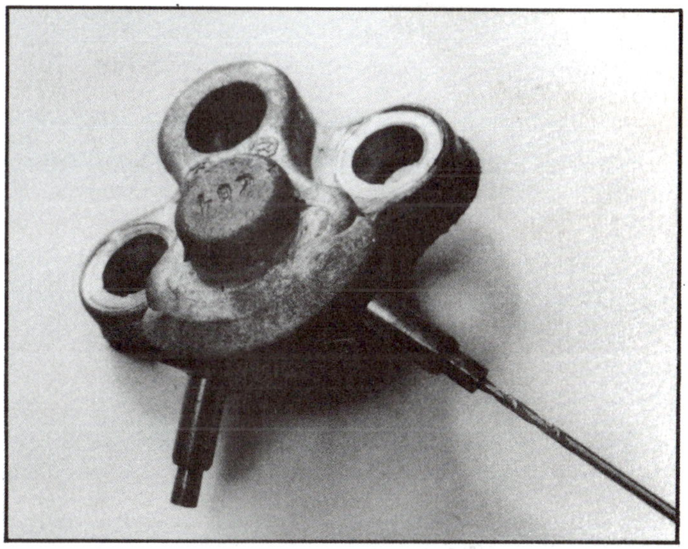

Diameter of the accelerator-pump discharge nozzle influences delivery rate. Drilling it to a larger size will allow a greater volume of fuel to be discharged earlier in the pump operating cycle and will often cure an "off-the-line stumble."

In a street-driven or road-race car, power enrichment timing is an important factor. On-and-off-again throttling requires a smooth transition from part-to-full-throttle enrichment; custom tailoring the metering rod piston springs may be necessary.

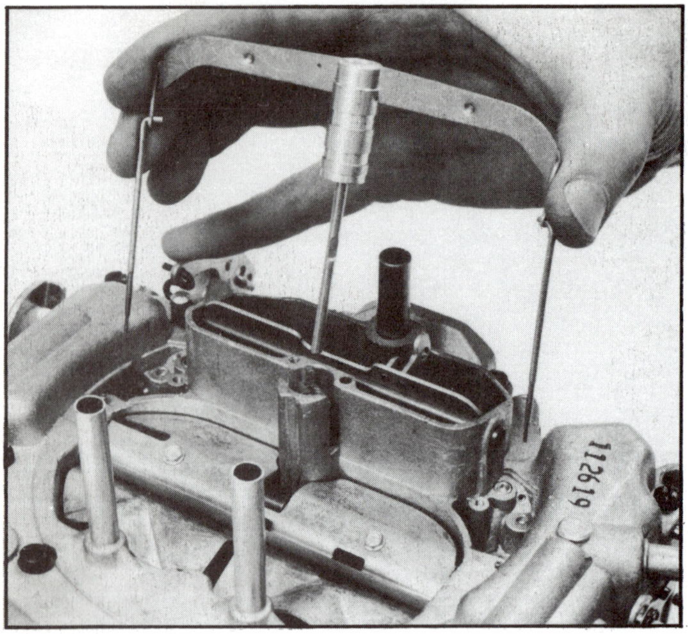

Metering rods offer a convenient means of recalibrating air/fuel ratio since they may be changed without disassembling the carb. In spite of the fact that many sizes of rods are available, it is also sometimes necessary to change jets in order to establish the desired calibration.

pump nozzle, only the *duration* of flow; as orifice size is increased, the same amount of fuel is squirted in a shorter time. While it might seem that larger is unequivocally better, it is possible to impair performance by admitting fuel too quickly. If nozzle diameter is too large, a cloud of black smoke will appear as the car leaves the starting line. This may very well be followed by a stumble as the engine encounters a momentary lean condition, occurring when the pump shot is dissipated before adequate fuel emulsion begins to flow through the main metering circuit. In order to avoid such a situation, start with the smallest discharge nozzle orifice and work towards the larger sizes. Optimum performance will be obtained by using the smallest diameter squirter that allows the car to cleanly leave the starting line. Once beyond this point, a quicker discharge rate is counter productive as fuel mixture will momentarily become excessively rich.

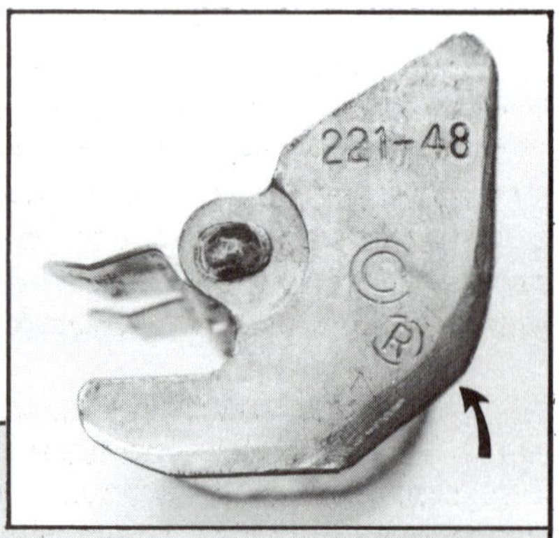

Quickening the opening rate of the AFB secondary airvalve is a matter of removing material from the counterweight. The upper weight has been modified (inset), bottom is stock.

POWER ENRICHMENT

A certain amount of time is required to get any fluid moving. Consequently, activation of the power enrichment circuit must be timed to avoid a transient lean condition as the engine is brought from part-throttle/no-load to full throttle/full-load operation. This is where the metering-rod-piston springs come into play. The stiffness of these springs determines the manifold vacuum level at which the metering rods are raised to the power step. And in a race engine it is important to adjust spring pressure so that enrichment is not engaged during idle, a common problem encountered with the low idle vacuum caused by racing camshaft profiles.

Carter does not correlate spring pressure to inches of Mercury but this can be accomplished with an easy test. Using a rubber hose or tapered fitting, connect a hand-held vacuum pump (such as the Mity Vac offered by Neward Enterprises) to the channel through which manifold vacuum reaches the cylinder in which the power piston resides. Then remove the metering rod chamber cover, work the pump, and note the vacuum reading at which the metering rod begins rising from its full-down position. This is the point at which the economy-to-power transition occurs.

Once this is determined, connect a vacuum gauge to the engine and check idle manifold vacuum. In order to avoid an overly rich condition, the idle vacuum should be greater than the vacuum required to hold the pistons in their down position. By way of example—if an engine idles at 12 in/Hg, power circuit activation should not occur until manifold vacuum drops to approximately 10 in/Hg. (It's best to leave a 1-1/2 to 2 in/Hg margin to compensate for gauge inconsistencies.)

SECONDARY TIMING

With optimized fuel circuit calibrations it is entirely possible that a faster secondary air valve opening rate may be tolerated. As we have mentioned earlier, the spring loaded valve used on AVS and Thermo-Quad carbs is easily adjusted. The AFB presents a different situation in that opening rate is altered by removing material from the counterweight. When introducing the AFB air valve weights to a grinder, proceed judiciously—it's much easier to remove than replace weight material.

CHAPTER 6 CARBURETOR TUNING

GENERAL TUNING TIPS

DIRT

I was once preparing to rebuild a carburetor, and in order to provide a proper environment in which to perform the disassembly and reassembly, I was carefully cleaning a well-used work bench. Puzzled by my sudden and uncharacteristic fervor for absolutely sanitary conditions, a bystander who was observing the spectacle blessed me with a sample of his sarcastic wit. I simply stated, "cleanliness is next to Godliness." "Even in carburetors?" "Especially in carburetors!"

Dirt is to carburetors as Professor Moriarty was to Sherlock Holmes. A good quality high-capacity filter should always be installed in the fuel delivery system. The Carter GF614 (left) and the Fram HPG-1 (right) are low restriction filters with replaceable elements that are available in many speed shops.

Too frequently, people who should know better lose sight of the fact that dirt and other foreign matter can wreak havoc with an otherwise perfectly adequate fuel system. Racers are especially vulnerable to dirt-induced problems since gas tanks from wrecked cars are often used, or a "custom" gas tank is made using improper materials or techniques. It seems strange that a drag-race car driven 200 miles a year has fuel system difficulties that never encountered with a street driven vehicle. To prevent unnecessary dirt-provoked fuel problems:

1) Always run a filter of some sort, even if it is just a brass Moraine-type element in the fuel bowl inlet. For high flow applications, use a universal high-capacity inline filter, Fram HPG-1 or Carter GF614; all have replaceable elements.

2) Inspect the vehicle gas tank periodically to

insure that internal coatings or galvanizing have not begun to come off.

3) When using electric fuel pumps, inspect the pump periodically for accumulations of dirt and/or rust.

4) Flush all fuel lines before installation or after rework.

5) If you install a fuel-flow measuring device (as is usually included in an onboard gas mileage computer) insure that it is of sufficient flow capacity to accommodate full throttle fuel requirements.

6) When allowing a car to sit for extended periods of time, flush tank and lines before refilling carburetor float bowls. Drain the gas tank and check for water in the fuel.

7) Inspect electric fuel pumps for clogging or jamming when they are unused for extended periods of time.

8) Add a can of "dry gas" at least twice a year to eliminate water which may ultimately cause fuel system corrosion.

9) Inspect the air filter periodically and replace as necessary. A dirty filter reduces economy and performance.

10) Spray all air bleeds periodically with a carburetor cleaner—such as Gumout—to remove dirt and fuel gum accumulations.

A maze of plumbing is Detroit's gift provided with every new car. Although satisfying emission requirements, this hose menagerie is a vacuum leak just waiting to happen. Rubber hoses can harden where they are attached, crack, and leak air without give any visual signs. A periodic check (remove the hose and inspect the ends) provides good preventative maintenance.

HEAT

The dichotomy of carburetor operation is that fuel should be cool when in a liquid state but heat is necessary for satisfactory vaporization. To this end, several techniques are commonly employed to keep the carburetor as cool as possible, while heat is applied either to the intake manifold, incoming air, or both.

For racing applications, it is common practice to block the heat riser passages in the intake manifold or use a manifold and/or cylinder heads with no plenum-heat provision. The "cooling" effect of this practice can improve horsepower by a few percent, and is useful in competition where every little bit helps. But for street driven applications, the blocked heat riser can be of more harm than benefit. The horsepower gain is still there to be had, but at the potential cost of fuel economy and driveability. Without exhaust heat being applied to the manifold, fuel vaporization is inadequate until the engine reaches operating temperature. In addition to operational problems, (which are not really all that difficult to live with) excessive amounts of liquid fuel in the intake manifold can reach the cylinder walls and wash away the oil that keeps rings and pistons happy. This may not be a problem in a race engine where throttle movement is minimal when the engine is cold. But with the "on-off-on" cycles that are required in traffic, or pulling away from stop signs or lights, the frequent contribution of fuel—from the accelerator pump—can lead to rapid ring wear.

In many late model vehicles, the incoming air is also pre-heated by the exhaust system. This is a further aid to vaporization and enables an engine to run well on a leaner air/fuel mixture for emissions

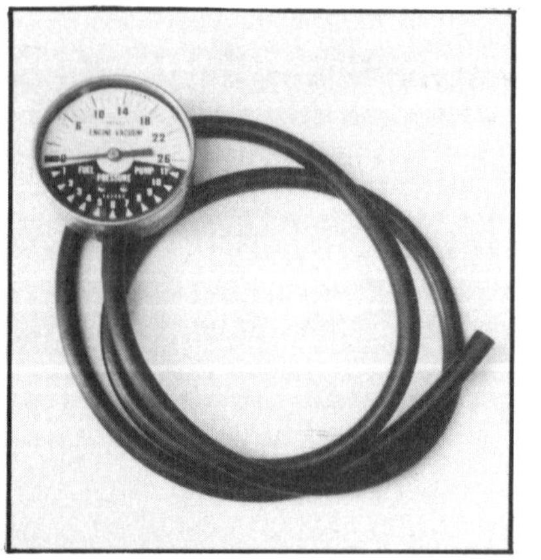

Tuning a carburetor without the help of a vacuum gauge is like flying blind; you can only guess what's going on. A vacuum gauge can not only help set proper idle and timing, but also it can be a valuable resource in detecting and tracing down vacuum leaks and equipment malfunctions.

Heat is both friend and foe to a carburetor. A little is needed for good vaporization, but too much can lead to poor performance, lean mixtures, and percolation problems.

Fuel economy, and good performance for that matter, is only obtainable when air restriction to the carburetor is minimized. The highly respected K&N cotton-wire element air filter (right) will usually outflow the standard paper design. The K&N uses a spray-on oil based fluid that permits cleaning and reuse.

When optimum fuel economy is the order of the day, appropriate tuning procedures should be followed and considerable thought given to such things as tires. Radials offer less rolling resistance than bias-ply tires and can measurably improve fuel economy.

purposes. The pre-heating systems are frequently discarded with no noticeable degradation in engine operation. Some systems are designed to assist in meeting cold-start emissions requirements, and as such have little positive effect on driveability.

Ideally, an engine should receive warm air when in cruise mode, and cold air when under full power. A high-performance engine with a vacuum controlled cold air intake would operate under just such an arrangement. In order to achieve optimum performance, it would be necessary to calibrate main metering for the warm underhood air (leaner), and increase flow through the power enrichment circuit (smaller diameter power step on the metering rods) to supply the fuel necessary for the cold air at full power.

Along with whatever air supply system is used, it may be desirable to cool the carburetor to minimize fuel percolation (fuel boiling in the float bowl resulting in stalling and poor hot-starting characteristics) caused by high engine compartment temperatures. A thick insulating gasket between carb and manifold will help, but if percolation persists, it may be necessary to install ducts or diffusers to deliver a flow of air past the carb exterior.

COLD

With all the plumbing that festoons emissions-era engine compartments, problems caused by extreme cold are rare. However when temperatures are between 30° and 50°F and humidity is high, carburetor icing can still occur during engine warm up. Icing of an automotive carburetor is almost always associated with low engine speeds (idle) where the gap between the throttle plate and bore is minimal. Water vapor in the air condenses, freezes, and blocks the gap, causing the engine to stall. Once normal operating temperature is attained, there is sufficient heat at the plate/bore junction to prevent icing. With a properly functioning choke mechanism (with fast idle provision) and the availability of manifold heat, icing should not be a problem.

STREET TUNING

TUNING FOR FUEL ECONOMY

When the price of gasoline broke through the $1.00 per gallon level, the term "fuel economy" took on new significance. And as the price of gasoline ebbs and flows, so too does public interest in fuel efficiency. Depending upon the current situation, gas mileage may not be the hot topic it once was, but neither is it treated as nonchalantly as it was when the pumps read 30 cents per gallon. Irrespective of public opinion, it makes good sense, economically and ecologically, to tune an engine to deliver optimum fuel economy, especially when performance does not have to be severely compromised. Carburetor tuning alone can do only so much—the key to hitting high mileage numbers lies in the overall engine combination, consisting of carburetor, intake manifold, exhaust system, camshaft and ignition. By properly coordinating these items, it is possible—albeit demanding—to achieve over 20 miles per gallon in a vehicle that is also capable of recording quarter-mile elapsed times of under 13 seconds.

The basic premise behind both efficient and high-performance engine operation is that a high output engine can loaf under many load conditions that would strain an under-powered powerplant. The situation is analogous to a 98-pound weakling, as opposed to a 200-pound athlete, attempting to lift a 150-pound barbell. Since the powerful engine is rarely stressed, manifold vacuum is consistently high, providing for efficient fuel vaporization and infrequent activation of the power enrichment system. A significant impediment to this approach, however, is exhaust emissions equipment, especially those systems fitted during the mid-1970's. Injecting exhaust gas, although good for emissions, will take its toll in terms of fuel economy and performance. But depending upon local laws, it may be possible to eliminate some items. Nevertheless, even if the full complement of pollution controls is retained, impressive overall mileage figures can be obtained with the proper techniques.

IGNITION REQUIREMENTS

Regardless of make, year or model, one of the first steps down the road to improved fuel economy should lead directly to a distributor machine. Most standard centrifugal-advance curves used by the automakers are designed to accommodate a wide range of drivers, including those who will lug an engine rather than down-shifting. However, an overly ambitious spark curve will induce detonation under some operating conditions. But with a little attentiveness on the part of the driver, initial advance may be moved to 14-16 degrees before top dead center (BTDC) and the centrifugal curve may be modified to provied an additional 20-22 degrees of advance at 2,500 engine RPM. In all, total mechanical advance (initial plus centrifugal) should equal 36-38 crankshaft degrees. The actual curve and total setting for each engine may vary slightly, depending upon combustion chamber shape, compression ratio, connecting rod length and vehicle weight, but the basic premise is to give the engine as much advance as it will tolerate without detonating, or as many old-time racers have stated: "Crank it up until it talks to ya, then back it down a little."

In addition to centrifugal spark timing, vacuum advance must also be included in the overall picture for optimum efficiency under cruise conditions. At one time it was common practice to disconnect and plug the vacuum line whenever a quick centrifugal curve was installed. This was done because most vacuum cannisters started moving the spark at comparatively low vacuum readings, and when

Substituting an electrical fan for the standard belt-driven mechanical model results in less parasitic engine load. The result—better fuel economy.

High energy ignitions such as the multi-spark unit offered by MSD Ignitions exert a positive influence on both performance and economy. Increased energy being discharged across the plug gaps will ignite lean mixtures that may cause a mis-fire with standard ignition.

combined with a modified centrifugal curve, part-throttle ping resulted. But in terms of fuel efficiency, overlooking the advantages of a vacuum advance system can be costly, decreasing mileage by up to four miles per gallon. The best approach is to tailor the vacuum curve, in essence delaying it so that an over-advance situation does not arise during the transition from cruise to power. Crane Cams and Cam Dynamics offer adjustable vacuum advance units for Chrysler, Ford, GM HEI, and point-type distributors. With this device, the vacuum advance curve can be easily adjusted so that the mileage advantages of light-throttle advance may be enjoyed without encountering detonation problems.

Finding an adjustable vacuum advance distributor for some engines is impossible, in these cases a bleed off may be inserted in the vacuum line to delay vacuum advance activation. This may be accomplished with a plastic T-fitting, a short length of rubber hose, and a carburetor jet. First, install the T-fitting in the vacuum advance line. Next, press the jet (.023-inch is a good starting point) into one end of the short rubber hose and install the other end on the T-fitting. The jet will bleed off some vacuum signal and delay the advance start-up until manifold vacuum is

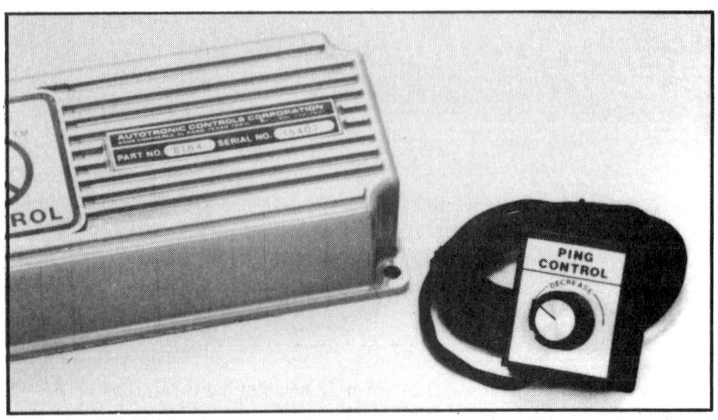

By allowing the driver to adjust spark lead from the dash, the MSD Ping Control offers a means of optimizing ignition timing for any driving situation, or for those occasional "low octane" tankfuls.

Take a good look at this engine. There are at least four obvious items that can be altered to improve fuel economy. What are they? 1) optimizing ignition timing and incorporating a water-injection system; 2) free-flowing exhaust headers and mufflers; 3) addition of high-torque intake manifold; 4) adding a high-energy, multiple-spark ignition system).

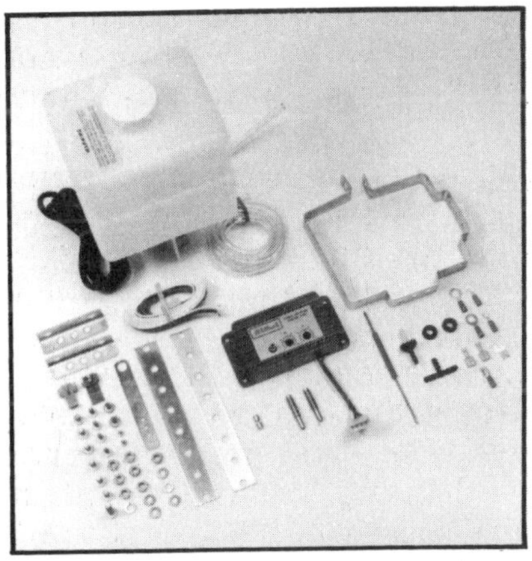

Water injection does not, in and of itself, improve power or economy. However, by quelling detonation, it allows the use of a more advanced ignition timing, and this can improve fuel economy.

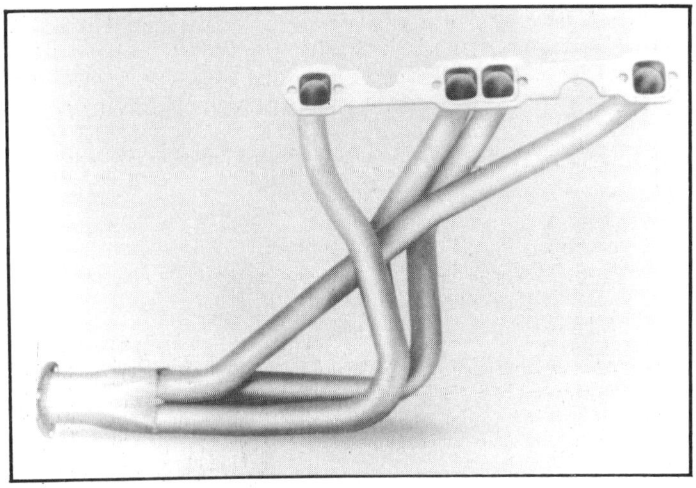

Exhaust headers have long been recognized as an important piece of efficiency increasing equipment. New technology such as these Cyclone anti-reversion headers, which incorporate a "cone" into the inlet to the header to cancel unwanted reversion waves, has generated even better results.

about 9 to 10 inches Hg. A larger jet will further delay activation, while a smaller orifice diameter will initiate vacuum advance at a lower manifold vacuum; try to tune for maximum advance without ping. (Note: if the advance line is connected to a straight, non-ported, vacuum source, a small vacuum leak, at the bleed jet, may occur at idle.)

As a final ignition modification, a high energy electronic ignition, such as the Autotronic Controls MSD 6-A multistrike box, should be installed if the engine is not already so equipped. The additional spark energy serves to eliminate misfires and also promotes more complete combustion, both of which contribute to increased mileage. Although the electronic ignitions that have been standard on new cars since 1974-75 are an improvement over the point-type systems, they too can be enhanced through the addition of a high performance ignition module—particularly a unit like the MSD ignition that generates multiple sparks per ignition cycle.

This is especially true of high-performance engines; the factory electronic ignitions aren't designed for high RPM (over 4,500) applications. Another approach to optimizing spark timing is to install an Autotronic Controls Ping Control. (The same company that produces MSD ignitions). The Ping Control unit allows the driver to alter spark timing simply by turning a dash mounted control knob. By so doing, spark lead can be retarded to eliminate ping during conditions of high load, and then advanced to optimize fuel economy during normal highway cruise. The ping control also serves to increase spark energy with stock ignition systems.

Whenever a high energy ignition is used, it is important to insure that the distributor cap, rotor and plug wires are in top-notch condition. Electricity always takes the path of least resistance, so if a distributor cap is cracked, or the wires have deteriorated, the spark may "leak" from the system and reduce ignition efficiency.

A LITTLE WATER

One problem that is beyond cure by any electronic ignition is detonation caused by a combination of enthusiastic spark timing and poor quality gasoline. It may well be that an ignition setting that optimizes fuel economy may cause the engine to knock and ping under full-throttle acceleration. An excellent solution to this problem was developed for piston-driven fighter aircraft during World War II: water injection. By introducing a carefully regulated volume of water into the air stream, combustion chamber heat, one of the prime causes of detonation, is reduced. Injection timing is accomplished by a vacuum switch that turns on a small electric water pump whenever manifold vacuum drops below a predetermined level. Therefore, at high load (low vacuum) conditions water is injected, ping is eliminated, and additional spark advance may be employed for improved economy under cruise conditions. Edelbrock and Spearco offer electronic water injection units.

AN EXHAUST SYSTEM FOR PERFORMANCE AND ECONOMY

Whenever efficiency within the combustion chamber is the goal, elimination of exhaust gases becomes critical. Headers have long been touted as mileage improvers, however, when fuel economy is the primary consideration, a decrease in primary tube size, as compared to performance/race style models, is in order. American Exhaust Industries, manufacturers of Cyclone and Blackjack headers, has done extensive research in the area of mileage headers, and has found that tubing diameters of 1-3/8 to 1-5/8 inches is most desirable. Understandably, high RPM performance is somewhat compromised, but at engine speeds of 5,000 RPM or less, the smaller diameter tubing typically represents either no loss in power, or a slight improvement, with a substantial increase in fuel efficiency over larger tube models. Another innovation is Cyclone's anti-reversion cone which boosts power and efficiency by minimizing the return to exhaust gases to the combustion chamber.

Of course, headers are of little help if the rest of the exhaust system is constrictive. For V-8 engines, dual exhaust pipes of 2 to 2-1/2 inches in diameter and a pair of free-flowing "turbo" style mufflers are de rigeur. It should be noted that all turbo mufflers do not have the same flow capacity, with several new designs having gone the traditional turbo muffler one better: improved flow with no increase in sound level. As an example, the Cyclone "Sonic Turbo" flows 82% as well as as a straight pipe, compared to other turbo mufflers that flow only 58%. The difference in flow capacity can have a significant impact on fuel economy, not to mention performance.

INTAKE MANIFOLDS

As with headers, high-performance intake manifolds have long been popular pieces of power-

In recent years, intake manifold designs have become increasingly sophisticated. Many street-oriented designs will optimize fuel economy and low-speed torque.

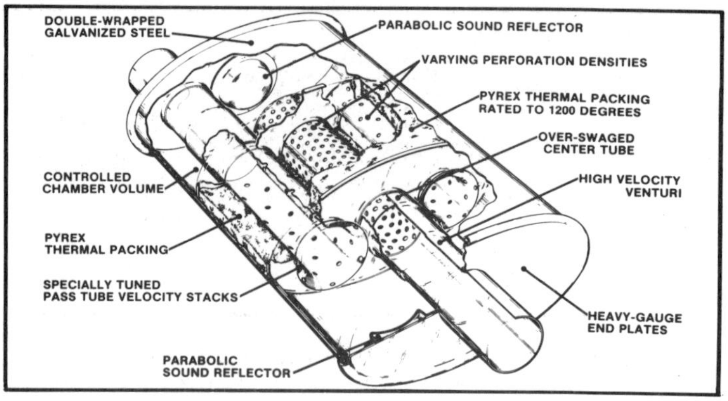

A low restriction muffler is an important complement to set of headers. The Sonic Turbo from American Exhaust Industries is one of the lowest restriction mufflers available that still provides acceptable sound control. Internally it differs substantially from conventional "turbo" style mufflers, which often contain little in the way of sound deadening and sometimes flow *less* than the Sonic Turbo.

increasing equipment. A number of years ago, the single-plane design was the rage, but aftermarket manufacturers have begun recommending the dual-plane configuration for many applications. In essence, the single-plane design was popular for high RPM, racing-type applications, and manifold companies varied runner size, length, and plenum size in an attempt to optimize performance within a particular RPM range. However, the single-plane design inherently produces less low-speed torque than a dual-plane configuration, causing a reduction in manifold vacuum at low engine speeds. In many instances, a plenum divider may be used in a single-plane manifold to improve low-speed response and intensify the vacuum signal at the carburetor. This will often improve driveability, but the increased

For general street driving, a good dual-plane high-rise manifold is hard to beat. This manifold is designed for 413-440 "RB" Chryslers and will accept either an AFB or Thermo-Quad.

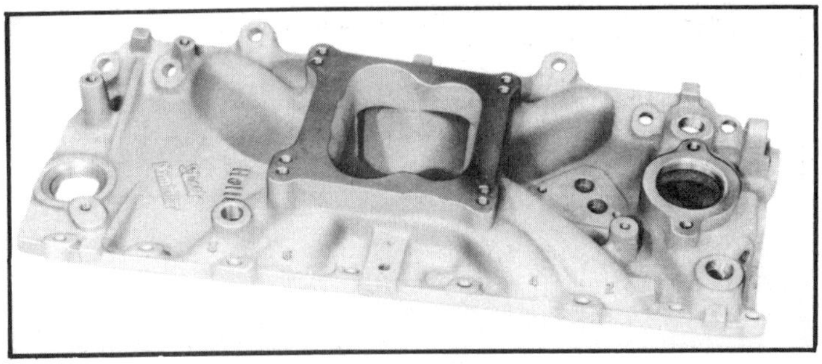

Although a single-plane open-plenum manifold may provide better fuel-consumption figures in steady-state dyno testing, it does not generally deliver outstanding low-speed torque. As a result, a single-plane manifold may decrease overall mileage in normal street use.

signal can lead to an over-rich condition, necessitating a change in jetting and idle calibration.

One manifold design, the Edelbrock SP-2P, embodies another approach in the quest for economy. A dual plane configuration, the SP-2P uses exceptionally small runners to increase the velocity of the mixture flowing into the combustion chamber. This helps keep the fuel atomized, but at the expense of power above 3,500 RPM. However, it has also been found that the SP-2P often does not work well in conjunction with other engine modifications, as it was designed for stock late-model powerplants with relatively mild cam timing, low compression ratios, and standard cast-iron exhaust manifolds. Weiand also offers a full line of dual-plane intake manifolds.

THE MAIN CONTROL ROOM

This again indicates the importance, but not the sole consideration, of coordinating components. The intended use of a vehicle plays an important role in determining which combination of equipment will prove most satisfactory. *Nowhere is this more important than in the area of camshaft selection.* One of the most commonly made mistakes in this regard is selecting a cam with excessive duration. Commonly called "over-camming," this practice results in a mismatch between the RPM range in which the cam was designed to provide optimum

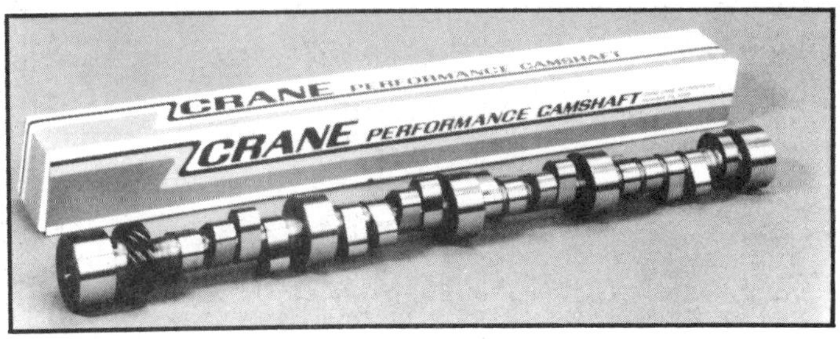

Camshaft technology has greatly improved in recent years. Like most cam manufacturers, Crane offers a number of profiles specifically designed to improve driveability and fuel efficiency. Always consult a reputable cam grinder/retailer before making a cam selection.

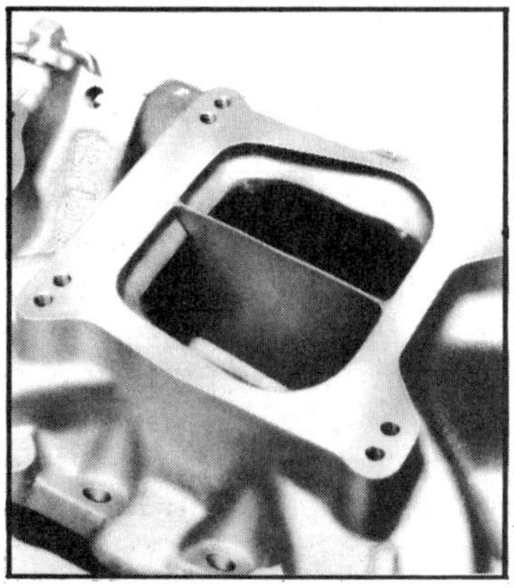

A plenum divider will improve the low-speed power of single-plane manifolds. The divider intensifies the fuel metering signal sensed by the carburetor, normally resulting in improved throttle response. The stronger signal, however, often requires rejetting if economy is to be maintained.

performance and efficiency and the speed at which the engine spends most of its time. According to Bob Stratman, performance consultant for Crane Cams, "We've found that the best way to select a cam is to determine engine RPM at 60 miles per hour and make your choice accordingly. Generally, as cam duration increases, so does the RPM range in which it was designed to operate. Therefore, we've matched our Econo-Power series of camshafts to specific operating conditions and compression ratios. Now our customers don't have to worry about all the technical stuff, like lift, duration, and overlap." Cam manufacturers do know what they're talking about, so it's best to consult with one if you're unsure about which profile to use. Many cam grinders, like Crane, offer a toll-free consultation line.

DRIVELINE CONSIDERATIONS

Manifolds, carburetors, camshafts, and headers—being intimately involved with engine operation—obviously exert an influence over fuel economy. But the drivetrain is remote. Its effects are not as readily apparent. Consequently, an engine may need to sip more than its ration of gasoline simply

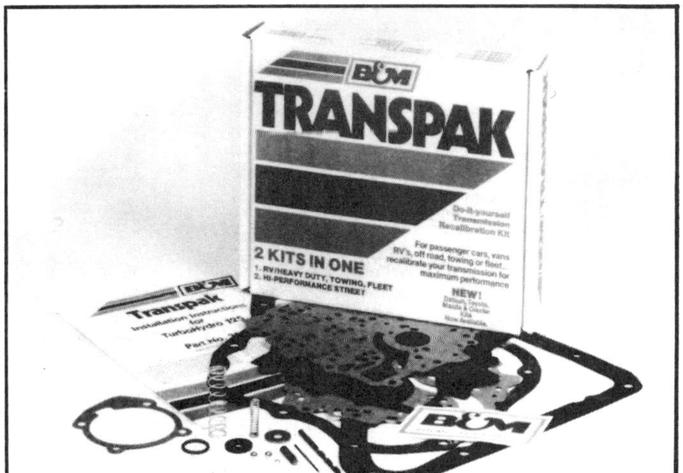

Slippage during shifting not only decreases automatic transmission life, it degrades fuel economy. Companies like B&M offer numerous kits that will provide shift recalibration and reduction of internal slippage.

because certain drivetrain components are inadequate.

A slipping torque converter and/or automatic transmission are the most common mileage pilferers. At best, these devices are 98% efficient in their ability to transmit power. With the passage of time, slippage can only increase, causing a resultant decrease in efficiency. The bottom line is an engine turning higher RPMs for a given road speed.

Based on typical gearing, an engine must turn approximately 1,000 revolutions per minute for each 25 miles per hour of vehicle speed. This translates to:

Vehicle Speed (MPH)	Engine Speed (RPM)
50	2,000
55	2,200
60	2,400
65	2,600
70	2,800
75	3,000

When the transmission is operating at 98% efficiency, this 2% loss manifests itself in the form of 44 additional RPM (for a total of 2,244 RPM) required by the engine to maintain a driveshaft speed of 2,200 RPM—the speed necessary for 55 miles per hour in high gear. However, when efficiency drops, engine speed must be significantly increased in order to maintain the desired speed, and that seriously impacts fuel economy.

% Efficiency	Engine RPM	Equivalent MPH for indicated engine RPM with no slippage
98	2,244	56
90	2,420	61
85	2,530	63
80	2,640	66
75	2,750	69

These efficiency calculations are theoretical with no consideration for tire slippage or other influencing factors, but they do establish a legitimate reference base. For example, with only 80% through-the-transmission efficiency, the engine is doing nearly as much work to propel the vehicle at 55 mph as it would have to do with no slippage to attain a 65 mph speed. Depending upon the vehicle, the extra engine speed can cost from one to four miles per gallon.

ECONOMY: AN OVERVIEW

Modifications can increase mileage, but frequently that's only part of the story—don't overlook common mileage thieves such as:

1) Choke—may be opening too slowly or never reaching full-open position. Bi-metal elements can deteriorate or hot-air tubes can become clogged or disconnected.

2) Air cleaners—when they become dirty, they can restrict air flow causing the mixture to richen.

3) Clogged air bleeds—can cause erratic operation and over-rich mixtures. Spray the carb throat with Gumout or a similar carb cleaner periodically.

4) Float level too hgh—contributes to an excessively rich mixture and possible fuel spill-over.

5) High idle speed—higher RPM increases fuel consumption.

6) Inoperative heat riser valve—if it's stuck open it will delay heating the intake manifold, reducing fuel atomization. If stuck closed, it restricts flow of exhaust gasses and can overheat the intake manifold and carburetor; may cause percolation.

7) Clogged muffler—interior of muffler can rust allowing baffles to come loose and restrict flow.

After all is said and done, there is no single "mileage motor" combination that will cure all ills. What is effective for one driver may not be for another. However, by following the general guidelines presented here and using trial-and-error test-

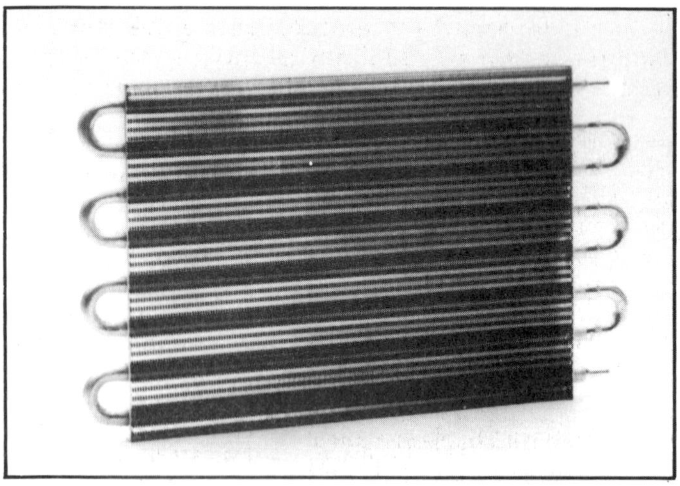

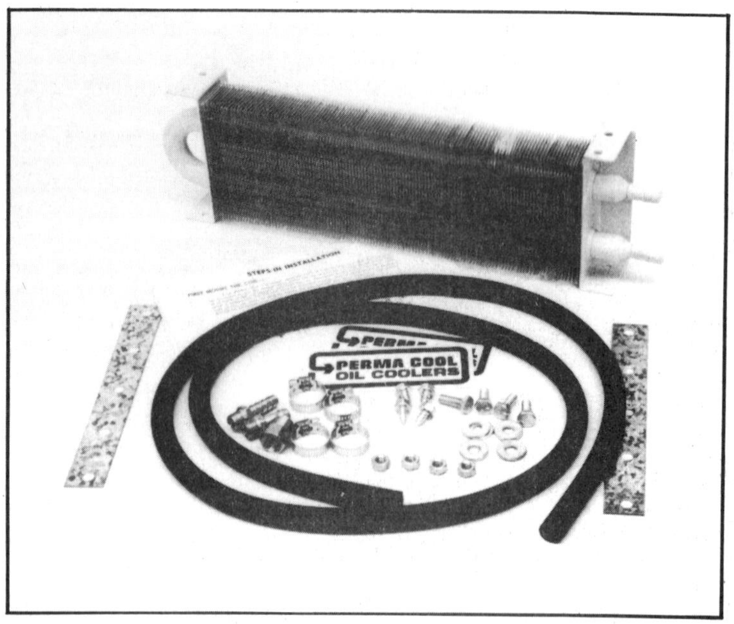

Excess transmission heat ultimately leads to increased slippage and outright failure. A cooler such as this one from Hayden (left) or PermaCool will keep temperatures down and extend trans life.

Some AFBs used on street vehicles have a tendency to stumble when the throttle is opened immediately following deceleration. If the stumble is caused by an excessively rich condition, drilling an air bleed in the primary boosters will alleviate the condition. The bleed should be drilled perpendicular to the angular face, as shown, using a 1/32-inch drill. The hole should be drilled into the passage just above the emulsion tube.

ing to fine tune, virtually any engine can be coaxed into delivering impressive mileage figures—but the driver has to do his part as well. If you constantly plant your right foot on the floorboard, you have to be prepared to pay the price. Erratic throttle operation is the most common practice that cuts into fuel economy. The best mileage may be achieved by holding the accelerator pedal steady and allowing speed to drop slightly when driving uphill. Each time the gas pedal is depressed, manifold vacuum drops (reducing fuel atomization), fuel is unnecessarily discharged from the accelerator pump, and the power enrichment circuit may be needlessly activated.

When driving for economy, the best practice is to moderately accelerate up to cruising speed, insure the transmission gets into high gear as soon as possible (without causing the engine to labor), and maintain a steady pressure on the throttle. One of the best ways to achieve this is to install a vacuum gauge and then simply adjust throttle pressure to achieve the highest reading. The vacuum gauge tells you how hard your engine is working. Simply stated: less work, less fuel!

TUNING FOR HIGH ALTITUDE

As a general rule-of-thumb, jet size should be *dropped one number (or .003-inch) for each 35° increase in temperature, or each 2,000 foot increase in altitude*, in order to maintain the same air/fuel calibration that was present at 70° and sea level. Since most of the inhabited portions of this country lie between sea level and 1,500 feet, where 70° temperatures are not uncommon, compensatory jetting for altitude or temperature extremes is generally not required.

However, for extended use at altitude, fuel economy and performance can be optimized by rejetting. Whenever this is done, somewhat lighter metering rod springs should be installed as engine vacuum also decreases as you drive to new heights. Each time the altimeter is bumped by 3,000 feet, power circuit activation should be delayed by 1.5- to 2-inches Hg.

Rejetting for a quick trip to the mountains is ill-advised. If the high altitude jetting is not corrected when returning to the lowlands, the excessively lean mixture could produce overheating and engine damage. Conversely, the opposite of an altitude increase occurs in an extremely cold climate. Cold, dense air may require richer jetting as a means of improving driveability during the long, hard winter. In either case, corrective action may be needed in order to satisfy local emissions regulations.

TUNING FOR TURBO AND SUPERCHARGING

When an engine is supercharged, either through exhaust driven or mechanical means, the intake tract becomes longer and cylinder pressures are

Adverse conditions warrant special tuning and fuel considerations. To keep this K&K kit car running on its blast up Pikes Peak, Bobby Unser used only the best air and fuel filters, plus a tuning expertise that made him famous.

Because they are very compact in relation to their air flow capacity, AFBs are used quite frequently in supercharger installations. This is a Paxton pressure box used in conjunction with a centrifugal blower. Pressurizing the carb necessitates replacement of brass floats with solid nitrophyl types.

increased. These conditions call for richer air/fuel mixtures and maximum obtainable accelerator pump discharge volume. The pressures involved in turbo and supercharging also dictate use of solid nitro phyl rather than hollow brass floats, as the hollow floats can collapse.

Due to its compactness and ability to survive backfires with little or no damage an AFB, or a pair of them, is usually the carburetor of choice on a Roots-type blower. Typically, primary main jetting is enrichened one step while secondary jet size is increased four steps. In conjunction with these changes, new primary metering rods, with a .010-inch to .012-inch smaller diameter (richer) power step are also installed. To ensure adequate fuel flow, it is wise to install .120-inch diameter needle-and-seat assemblies (part number 25-862).

On any draw-through blower installation, the carburetor does not directly "read" main body vacuum—it's on one side of the supercharger while the engine is on the other. Ideally, the power enrichment circuit should be externally referenced, meaning that the conventional power piston vacuum ports are blocked and replaced by new ones connected to the engine side of the blower through tubes. With an AFB, this requires use of a spacer which blocks the vacuum pickup passages from the inlet side of the blower and contains new channels which may be connected to external lines.

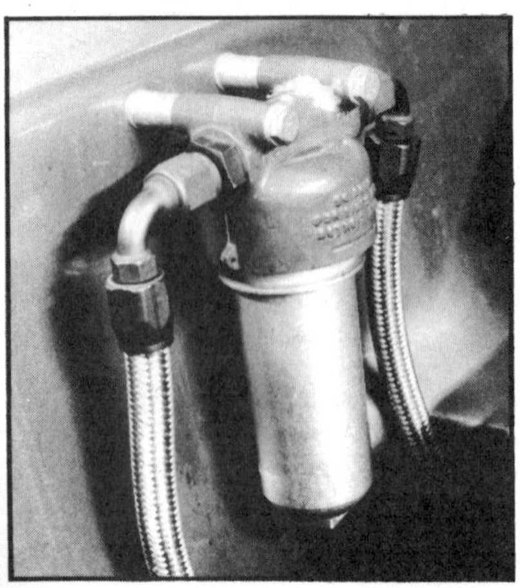

Replaceable-element filters, like this aircraft unit on the left or the Carter GF614 on the right, are the best choice for both on- and off-road applications, as they provide low restriction and can easily be serviced.

Dirt flying everywhere looks impressive, but looks can be deceiving. Since dirt and carburetors aren't known for their ability to mix, a good quality air filter is absolutely essential.

Since this modification is a bit involved and results in extra plumbing, another approach is frequently used. By using a stiffer metering rod piston spring (or stretching the existing one) the power circuit will be activated at a higher level of manifold vacuum. This partially compensates for the difference in vacuum between the inlet and outlet sides of the blower. An over-stiff spring will result in a mixture that is too rich rather than excessively lean—a move in the right direction. Lean mixtures and superchargers are as compatible as a plaid jacket and striped trousers, but the consequences of combining them are far more dramatic. Lean mixtures often result in serious damage to the engine, the blower or both. For that reason, a high-capacity fuel delivery system is called for and fuel levels in the float bowls should be adjusted slightly higher than normal.

TUNING FOR OFF-ROAD COMPETITION

There is a strange excitement in blazing new trails, especially when it requires chugging through rock-infested ground at 15mph one minute and charging across a dry lake at over 100 mph the next.

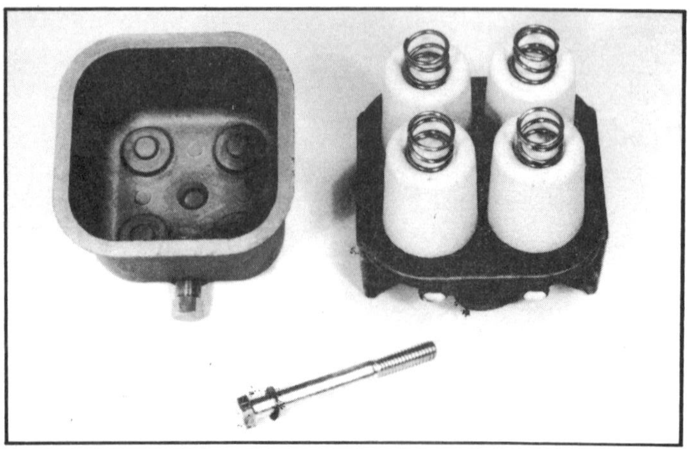

Such "activity" induces an effect in the carburetor, not unlike you or I attempting a sip of coffee during an earthquake. Wide RPM and load ranges teamed with unruly terrain, make consistent fuel metering no easy task. Carburetor modifications are definitely called for, yet they are surprisingly minor. Even for a full-bore racer, only a few dollars worth of parts and a Saturday afternoon will turn a milquetoast street carb into a back-woods super star.

The Carter AFB, Thermo-Quad and AVS four-barrels all use air valves for secondary operation and are therefore quite acceptable for off-road use. Mechanically-actuated secondaries should be avoided as erratic foot-throttle control can lead to an excess amount of horsepower being applied to the rear wheels. This on-and-off throttling results in excess wheel spin and a reduction or complete cessation in forward movement as the tires churn loose sand, gravel, or mud into the air. Air valve secondaries tend to open more slowly, and more smoothly, and are easier for the driver to control when he is bouncing around the cockpit.

The Carter 625-CFM 9000-Series AFB offers an excellent combination of low-speed driveability and high-speed power potential, and is probably the best choice for an off-road carb. Smaller engines may be better off with a 500 CFM model, but irrespective of the carb used, the standard needle-and-seat assemblies should be replaced with part number 25P-965S. The inlet-needle valve in this assembly is a two-piece affair with a small coil spring insulating one part from the other. The spring functions as a shock absorber, to dampen erratic float movement which results when the vehicle is driven briskly over rough terrain.

MAINTENANCE—ESPECIALLY IMPORTANT

As much as anything else, periodic maintenance

Replacing the air filter with a high-flow K&N element can improve both performance and mileage. This unique filter uses a cotton material that can be cleaned and reused many times. K&N filters are well known for their low restriction and superior filtering ability.

is essential to the well-being of any carburetor used in an off-road environment. The preponderance of dirt and dust dictates that fuel and air filters (absolute necessities) be checked regularly and replaced when necessary. A Carter GF614 or other high-capacity fuel filter should be used in conjunction with a K&N or similar high-flow air filter. Whenever the carburetor is inspected, the air bleeds should be examined to insure that they are free of foreign matter. A shot of Gumout carburetor cleaner will remove all but the most stubborn blockages. Just think of Gumout as Drano for carburetors. If it doesn't cure the problem, the use of a "plunger"—in the form of a fine wire—will be necessary. In most respects, a first-class off-road carburetor may be built using the same techniques employed to tailor an AFB, AVS or Thermo-Quad for street or race usage. Check the modification chapter for additional details.

AVOIDING VACUUM LEAKS

Vacuum leaks in the intake system are most frequently associated with manifold installations. But that doesn't preclude the possibility of a leak resulting from a carburetor swap. Fortunately, most carburetor induced vacuum leaks are of sufficient proportion that there can be no doubt as to their presence. An unplugged vacuum pick-up tube is easily located by virtue of the highly audible hiss caused by air being drawn into the opening. But if a ported vacuum hole is left open to the atmosphere, there will be no indication of a leak at idle; the additional air will be drawn in only at part-throttle, higher-RPM levels.

Some choke linkage mechanisms are incompatible with aftermarket manifold designs. Whenever a carburetor is to be installed for the first time, it should be placed in its normal position, and "rocked" to insure that it is not being held up by some carburetor protrusion or linkage arm. If interference exists, it is usually corrected by filing the outer edge of the mounting flange or removing the offending manifold material. But make sure you fix the interference problem *before you tighten the carb to the intake manifold*, or severe damage to the carb may result.

Another cause of vacuum leakage is over-tightening of the carburetor attachment nuts or bolts. This can cause a cracked throttle body, and frequently the crack will intersect a vacuum passage. Carburetor attachment hardware should be just snug enough to prevent engine vibration from causing the fasteners to loosen.

CHAPTER 7
REBUILDING TIPS

THERMO-QUAD REBUILDING TIPS

Rebuilding a Carter Thermo-Quad is not an especially difficult undertaking, however, a few words of caution will keep you from questioning your sanity once the carburetor has been disassembled. One important suggestion concerns removal of the accelerator pump from its chamber. As shown in the photo on page 82, a hammer and punch are used to remove the pump intake check assembly at the bottom of the chamber. What isn't obvious is the gusto with which the hammer must be used—a light tapping won't come close to doing the job. Be sure to position the punch squarely on the top of the pump rod so that it doesn't slip off and chip the air horn assembly.

One is also well advised to treat the phenolic resin main body with a bit of care. It is, after all, a hard plastic and although it is extremely durable, it can chip or crack if it is dropped or mistreated.

DISASSEMBLY

Before ripping things apart, give the carb a visual once over, being careful to note (and remember) the connection of various linkage pieces and the mounting of choke paraphernalia.

Nothing more exotic than garden-variety tools need be used in the assembly and disassembly of a Thermo-Quad. Not shown is the secondary air valve adjustment tool (part 109P-397). It's not absolutely essential, but it does make life easier. If a carburetor stand is unavailable, use a block of wood to elevate carb from work bench and prevent damage to linkage or throttle plates.

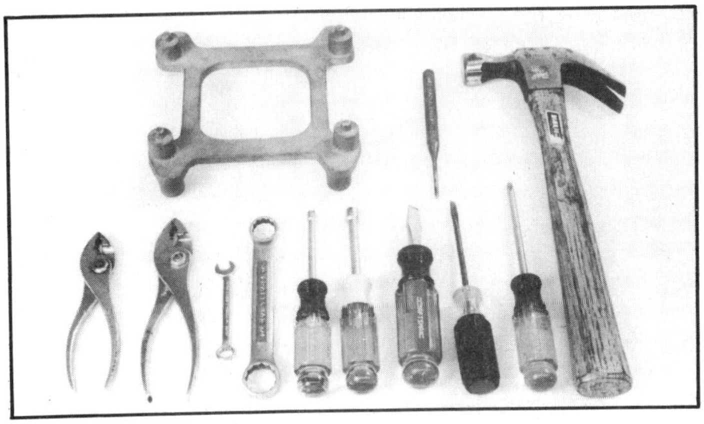

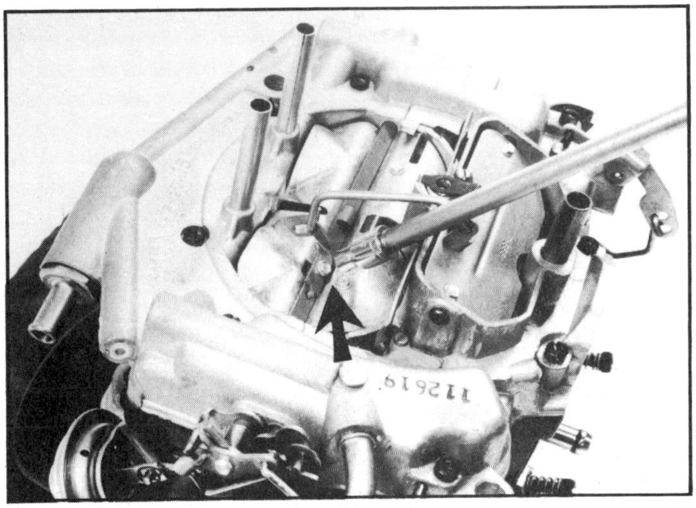

In order to remove the metering rods and hanger bracket, a portion of the secondary lock-out linkage must be detached from the transfer rod. This linkage prevents the secondary throttle plates from opening while the choke is closed.

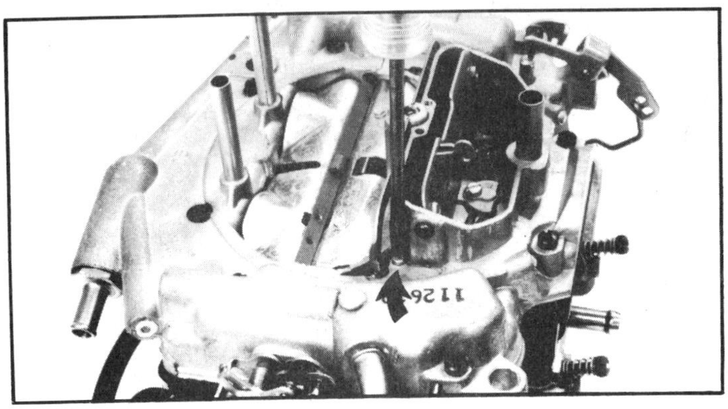

Remove the screws that attach the metering rod retaining plates to the air horn. There is a screw and plate on each side.

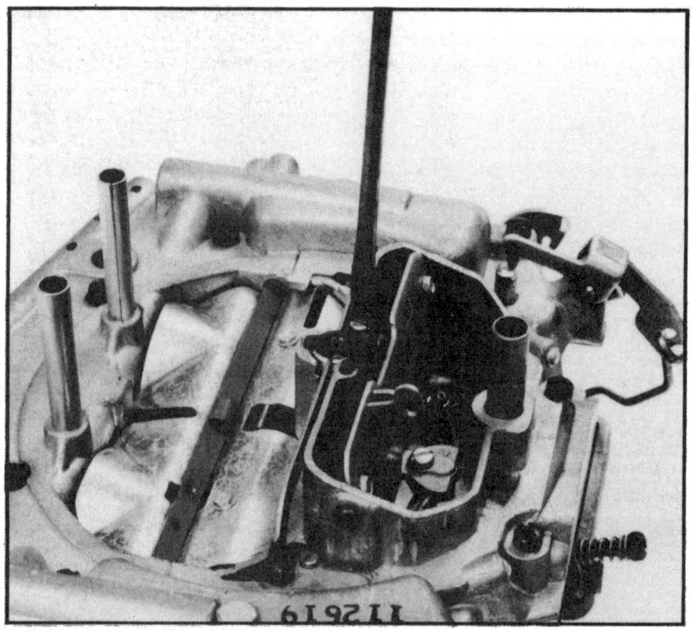

The metering rod hanger is held in place by this spring clip.

Once the hold-down screw and clip are removed, the hanger and rods may be pulled out. The metering rods are not held in the hanger bracket when the bracket is lifted out, so unless you grasp the rods when lifting the hanger bracket they will "self disassemble." Note the dimples in the hanger (arrow). The hanger should be installed so that the side with the two dimples faces the choke horn; they serve as alignment surfaces.

Next pull the spring from the metering rod piston housing.

A 3/4-inch box wrench or a deep socket can be used to remove the inlet fitting. This should be done before the carb is further disassembled, as it frequently requires a considerable amount of force to break the fitting loose; better leverage is afforded by gripping the entire carb, rather than attempting to grapple with just the air horn assembly. In fact, it's a good idea to break a stubborn inlet fitting loose while the carb is still bolted to the intake manifold.

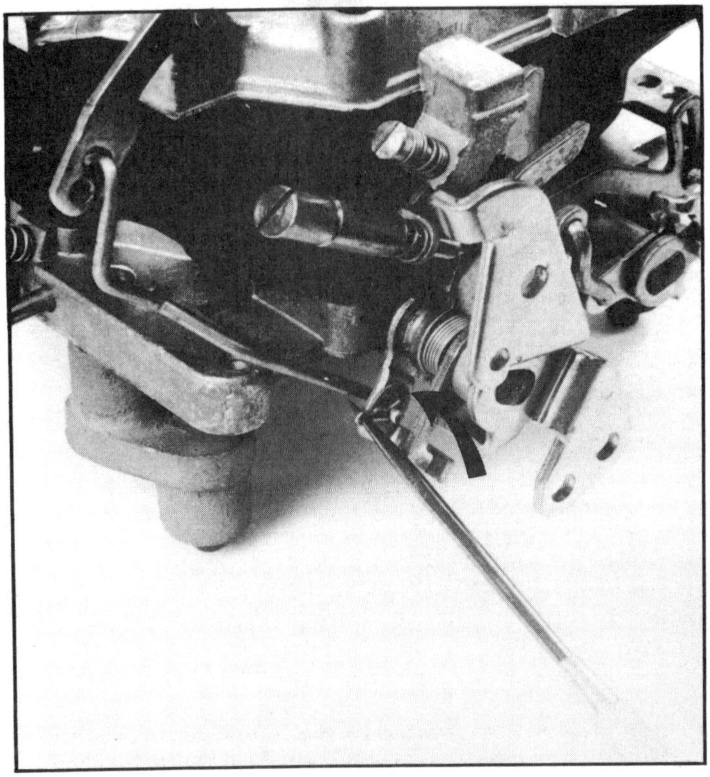

Remove the accelerator pump link rod by pulling off the retaining clip and slipping the rod out of the hole. Note that there is a washer behind the clip.

The pump rod is hooked on one end, so it need only be lifted and rotated to remove it from the pump lever. Note the hole in which the pump lever was located so it can be replaced in the same position upon reassembly.

The choke linkage is on the opposite side of the carburetor from the accelerator pump linkage. Remove the retaining clips at each end of the pull-off diaphragm link. Then remove the link rod.

Loosen and remove the screws that hold the the three main castings together. Since these pass from the air horn assembly through the phenolic main body and into the throttle plate, they are fairly long.

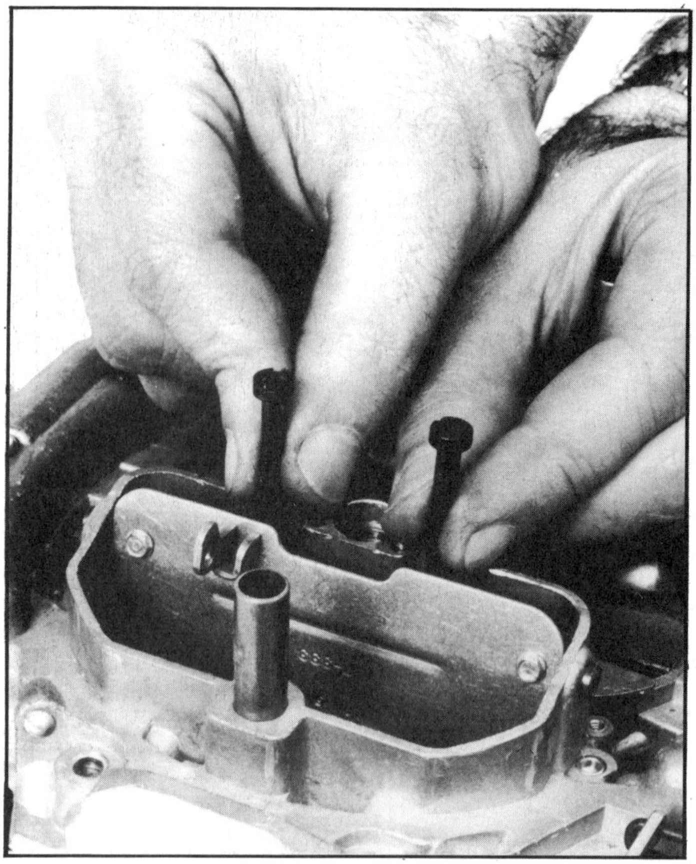

A total of ten screws are used, and all are the same length. It's easy to overlook the two that are down inside the choke horn.

The phenolic main body may now be lifted from the throttle plate assembly.

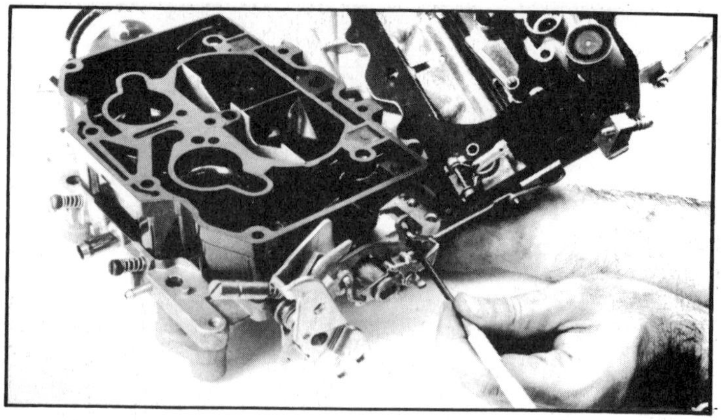

With all the screws, metering rods, and the linkages (almost) out of the way, the air horn assembly may be lifted and rotated towards the rear. The transfer rod-to-throttle lever link is hooked on the end and cannot be removed until the air horn is lifted and rotated.

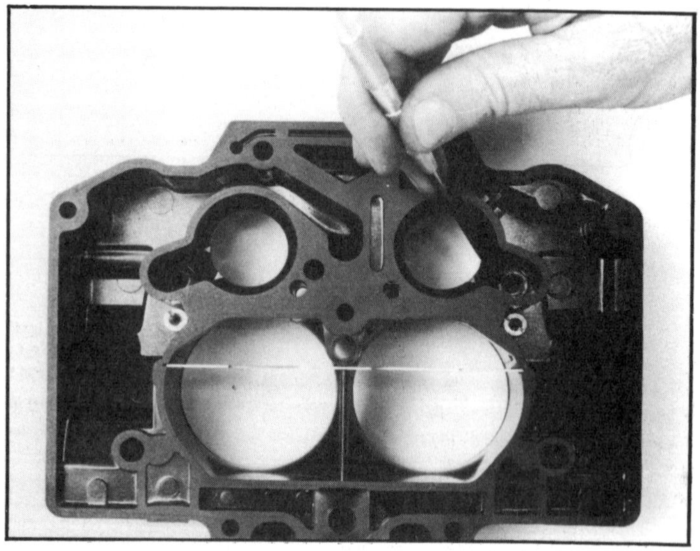

O-rings are used to seal the main well castings, which extend downward from the air horn assembly, to the main body. Always replace these seal rings; ignoring this advice can lead to unstable air-fuel ratios and fuel leaks.

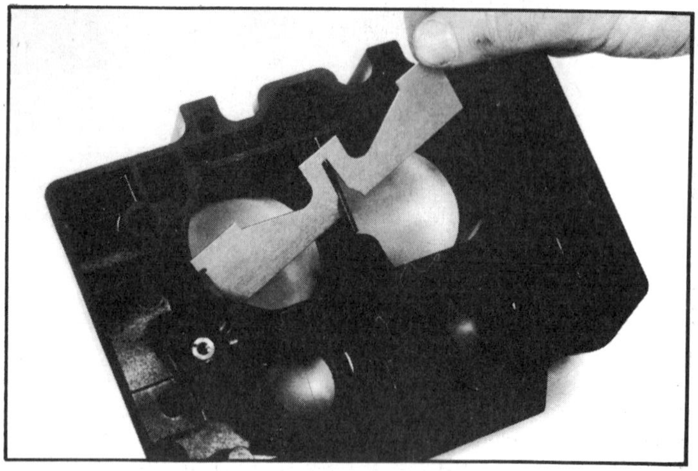

Secondary throttle bore diffuser should lift out easily.

Stripping of the throttle plate assembly is next. Begin by removing idle-mixture screws.

Here the step-up piston lever is removed.

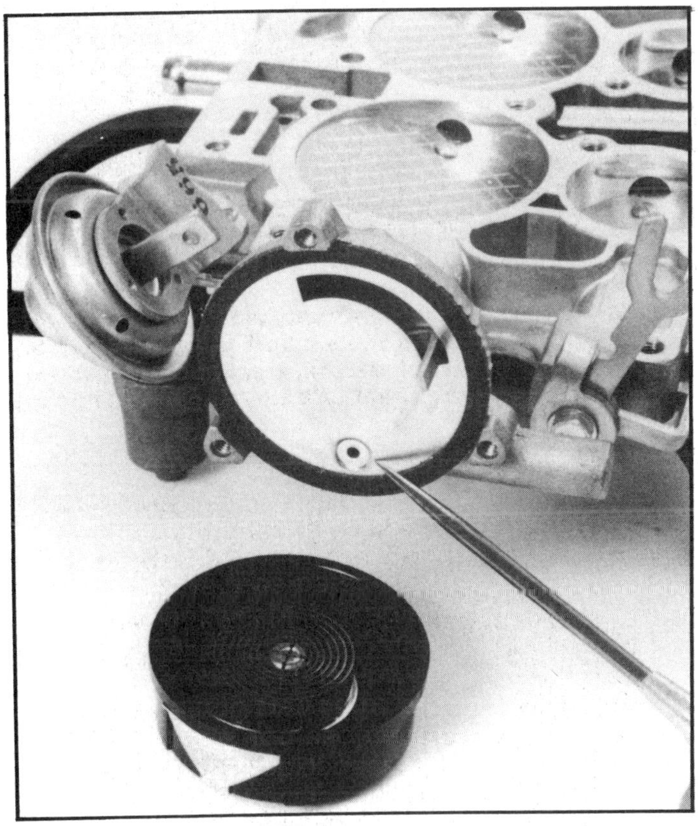

The choke housing is held in place by three screws and a retaining ring. Once they are removed, the housing may be pulled loose to reveal the backing plate. Note position of hole for reassembly. This may serve as an inlet for exhaust-heated air on some models.

The cast-aluminum portion of the housing is held to the base plate with two Phillips-head screws, one of which is recessed and easy to overlook.

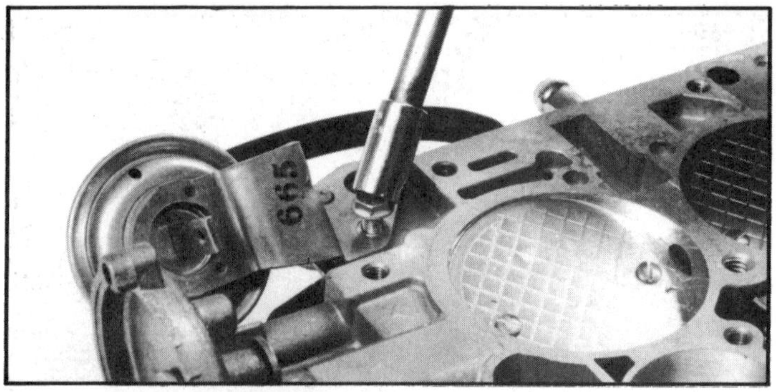

A 5/16-inch socket or open-end wrench will remove the screw that holds the choke pull-off diaphragm in place.

The air horn assembly contains many major components, including the floats. They may be removed by first lifting the gasket and coaxing the pivot pins loose. When the pins are free, the floats may be removed.

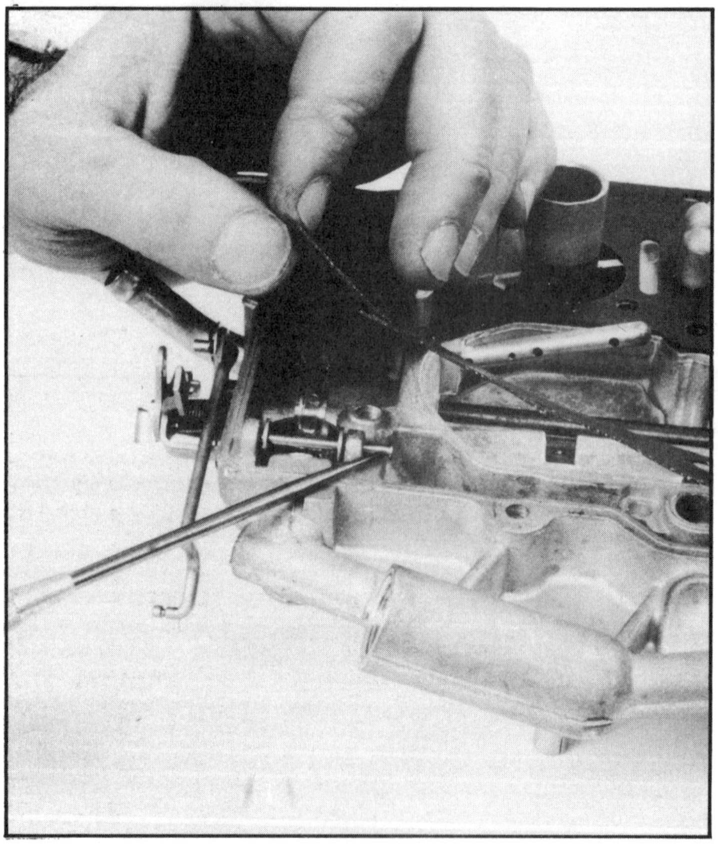

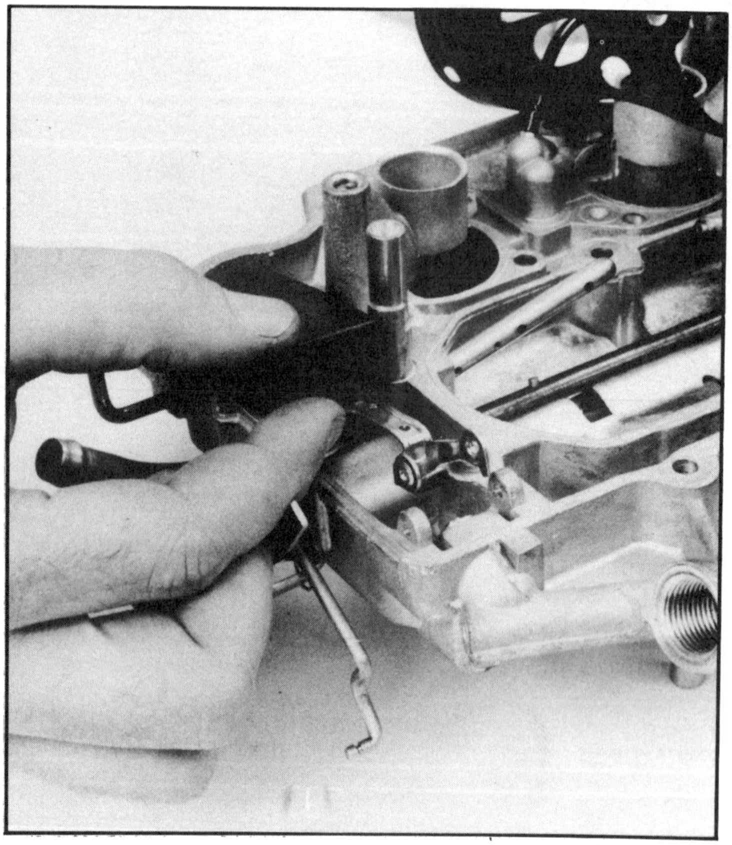

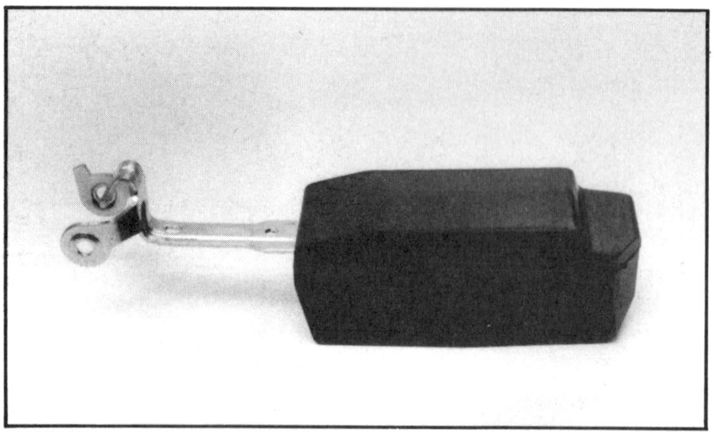

While the floats are away from their normal residence, it's a good idea to check them. All TQ's use float number 21-252, which are manufactured of nitro-phyl material. They are porous and will sometimes begin to absorb fuel, which causes them to sink rather than float. Weighing a float is the most accurate method of determining whether a float is still good. It should tip the scales at between 7.5 and 8 grams. A quick check method is to rest the float on its bottom surface and push downward at the pivot point. If it remains in this position it is good, but if the nose drops back down (direction of arrow), it should be replaced. The right photo indicates the way the float should balance if it is usable.

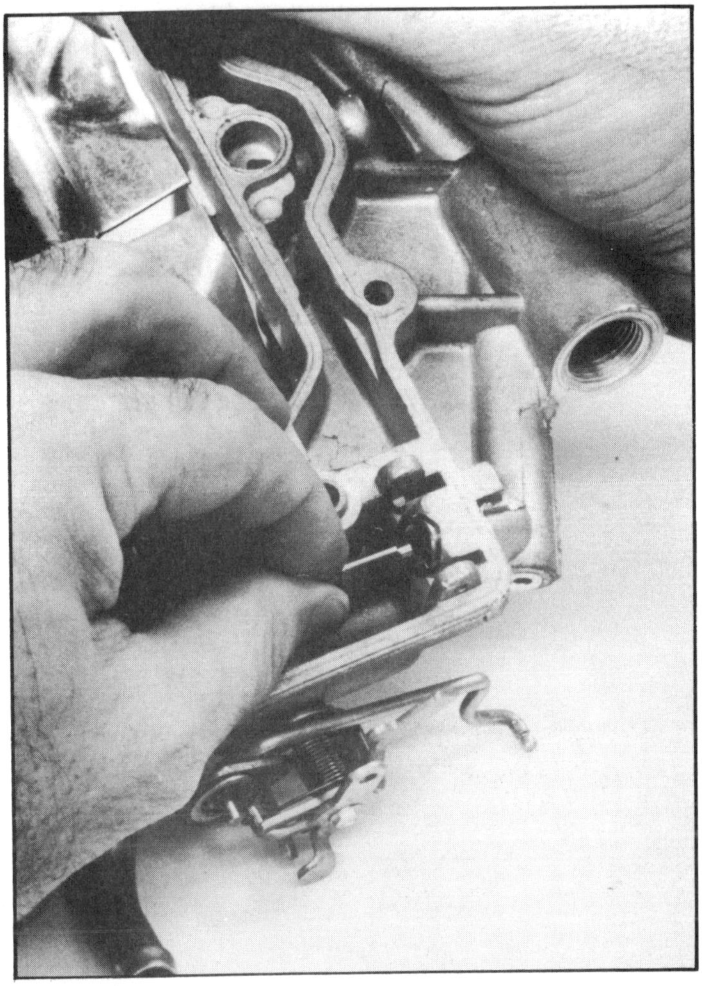

Once the float is removed, the inlet needle valve may be pulled from its seat.

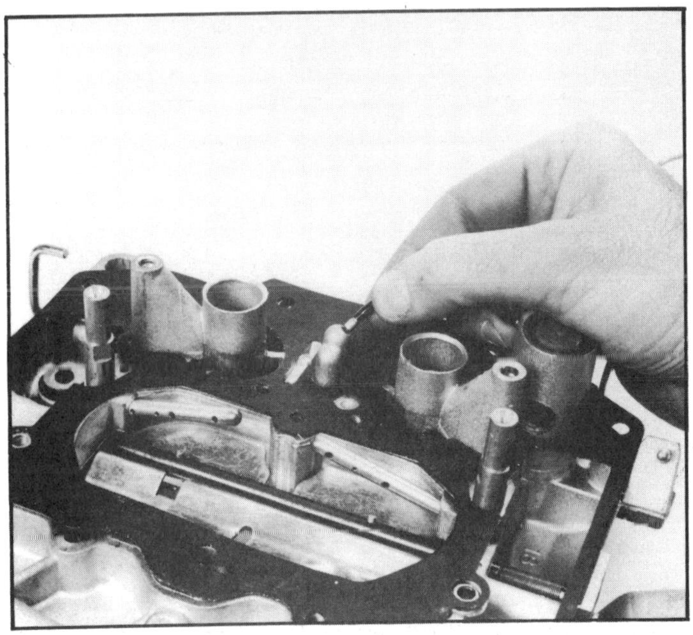

The accelerator pump shot travels through a plastic tube before reaching the discharge nozzle. The tube, and/or the fitting over which it slides, may be removed by a firm tug. With the tube pulled loose, the air horn gasket may be lifted off.

Then a wide-blade screwdriver is used to remove the seat. The gasket which fits beneath the seat should also be removed; it may sometimes stick to the airhorn assembly.

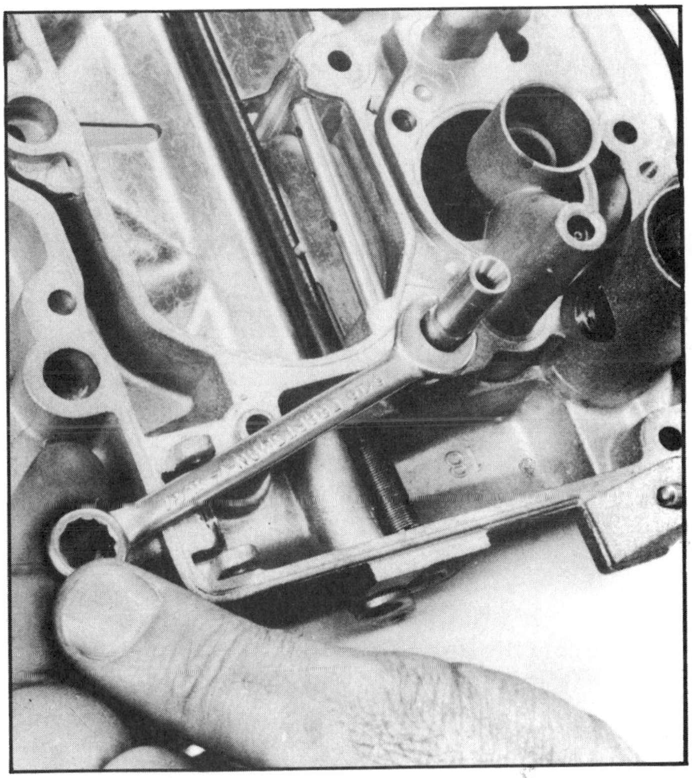

This is one of the secondary main jets. Two flats are milled on the outside surface, and a 5/16-inch open end wrench may be used to loosen it.

The accelerator pump discharge nozzle is held in place by a single screw.

Once the retaining screw is removed, the check needle will fall out if the air horn is inverted.

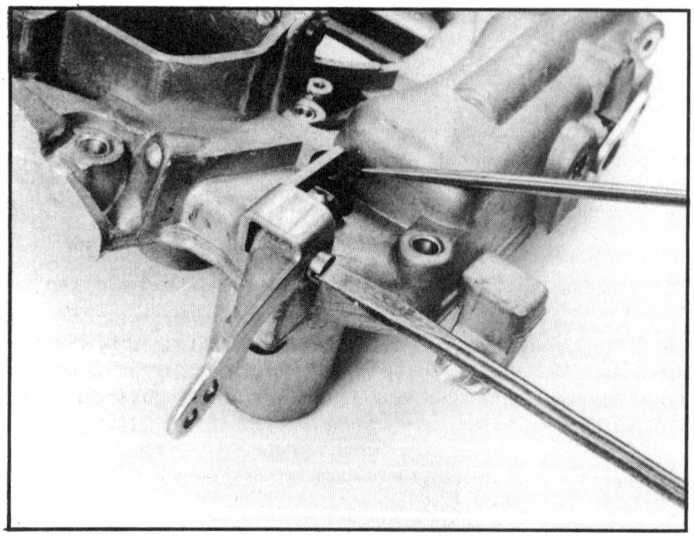

Removal of the accelerator pump plunger begins with the loosening of the the lever-pivot screw. Then the lever may be raised and slid off the "S" link.

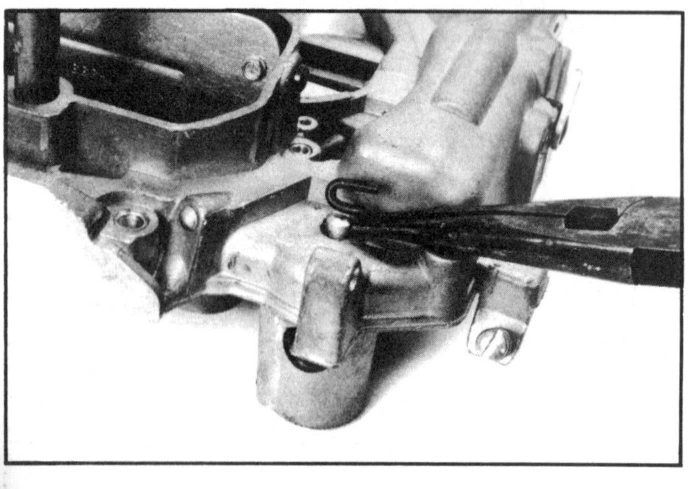

Pull the "S" link out of the plunger stem with needle-nose pliers.

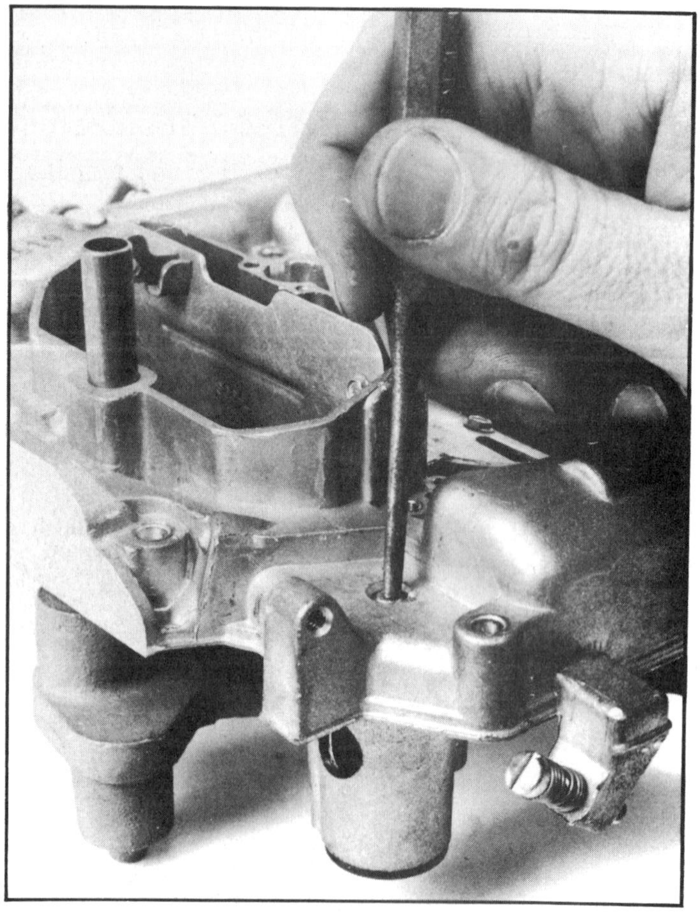

The stem will then be forced downward by internal spring pressure. At this point, the aforementioned hammer and punch must be activated to drive the intake check assembly out of the pump housing.

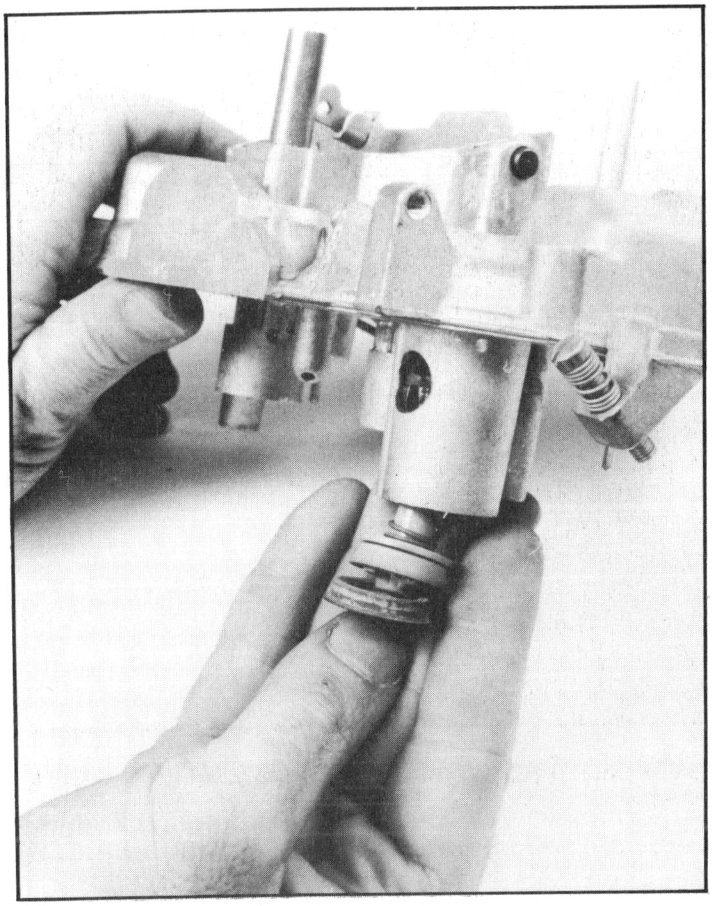

Voila!—after some hammering, the check assembly and plunger are removed.

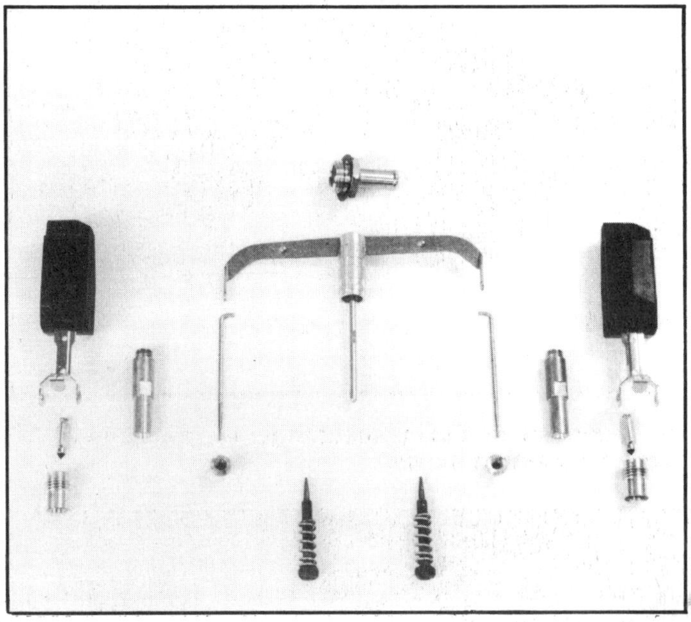

Before reassembling the carb, check all components that are to be reused and also verify that replacement parts match those that are to be discarded.

REASSEMBLY AND ADJUSTMENTS

Reassemble the carburetor by simply reversing the disassembly process. All components and subassemblies should fit together easily. If you have to apply force, something is wrong—determine what it is before proceeding. The only exception to this is the reinstallation of the accelerator pump check valve; it must be "drifted" into position with a correctly sized punch and careful (but forceful) hammer taps.

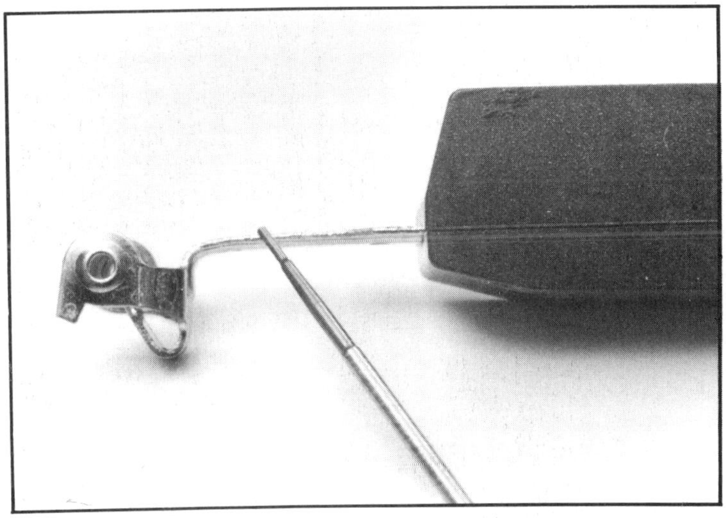

Float drop is not adjustable on Thermo-Quads. Level may be set by bending the float arm with a pair of pliers (at the point shown in the left photo) and measuring as shown (on the right). Rebuild kits or factory manuals contain float setting specifications for individual applications. For performance use, floats are generally set slightly higher in the bowl (a reduced measurement on the scale) than normal. This allows a greater volume of fuel in the bowl and reduces (slightly) the possibility of fuel starvation at high RPM.

Setting of secondary air valve preload for optimum performance is a trial-and-error proposition, which must be done with the carburetor installed on an engine. As a starting point, with all pressure released, turn the spring screw counter-clockwise until the valve just closes (make sure that there is no interference from any linkage). Then rotate the screw an additional 1-1/4 turns and lock it in this position. When making further adjustments, hold the adjusting screw while breaking the locking plug loose.

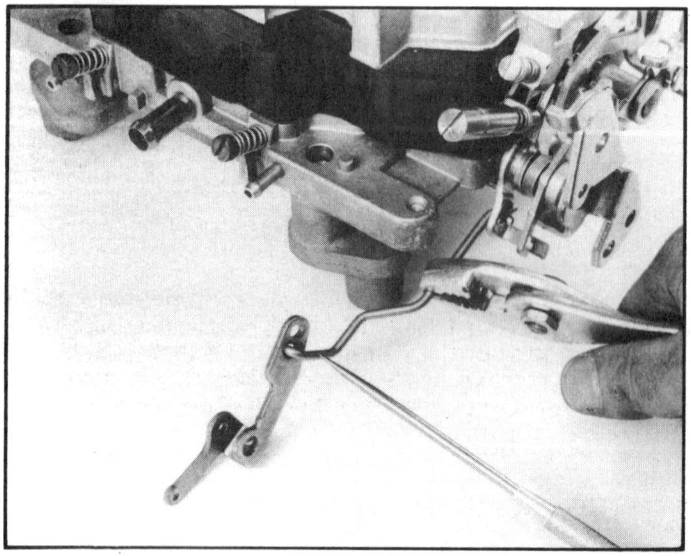

Accelerator pump travel is adjusted by bending the link rod. As the rod is "shortened" (less straight), pump travel and, consequently, discharge volume is increased. Compound bends in rod can cause it to hang up on base plate, so be careful. Always rotate the throttle shaft from idle to wide open and check for interference when making pump rod adjustments. Also note that hooked end of rod is in center position on lever arms with three holes, and in inner slot (as shown) on two-hole arms.

Generally, preload need be varied only slightly and the easiest method is to proceed from the current setting rather than releasing all pressure and starting for base zero. However, by noting the adjusting screw position, and the number of turns taken after valve closure, a setting can be easily duplicated once lost. At worst, you'll be 1/2 turn off.

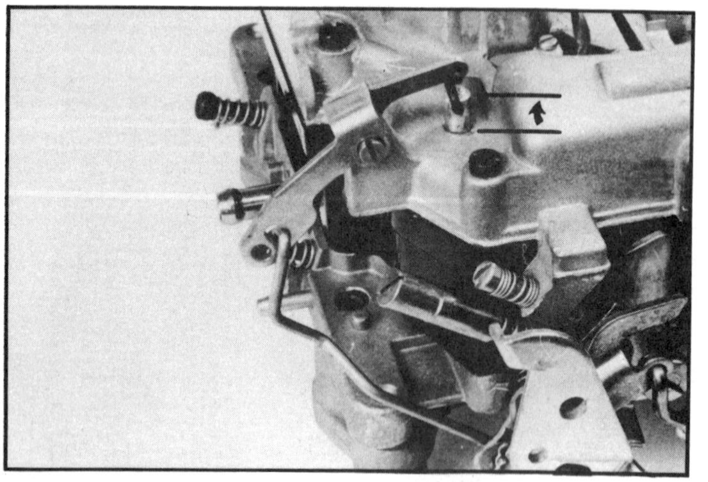

Many Thermo-Quad carbs have a "staged accelerator pump" that is set to provide a small additional shot when the secondary butterflies are opened. Adjustments are made with carb in the idle configuration and at the point where the secondary latch just makes contact. Factory manuals list the specific recommended dimension between the top of pump stem and the top of the air horn for individual carb models. Generally, the stem top should be approximately 17/32-inch above the air horn with the throttle shaft in the idle position. This dimension should be approximately 5/16-inch when the throttle lever contacts the secondary lock latch. For most performance applications, the pump is set to slightly larger dimensions for maximum fuel shot.

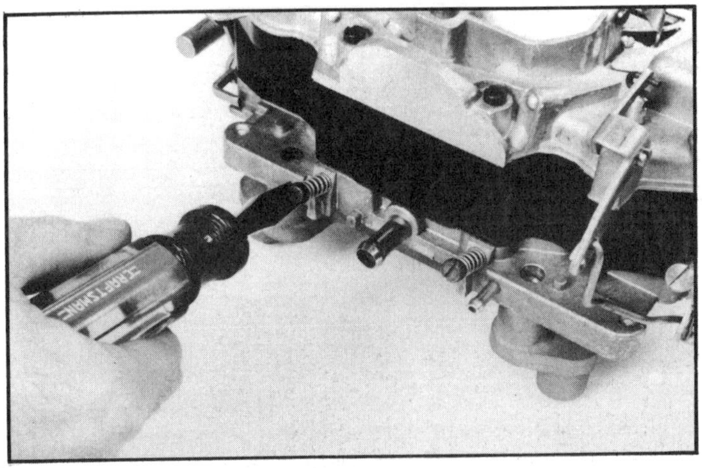

Final idle-mixture adjustments should be made with carb installed on engine. However, a starting point can be achieved by rotating the screws clockwise until they bottom, (do not force them) and then counter-clockwise 1-1/2 turns. Use a vacuum gauge or tachometer to make the final adjustments (adjust until highest reading is obtained).

With the engine cold, open the primary throttles and check to ensure that fast-idle adjusting screw rises so that it rests against the fast idle cam (arrow). Adjust the screw so that it holds the throttle open slightly.

After the mixture screws have been set, adjust engine idle speed. The lower screw in this photo controls fast idle when choke is applied.

To adjust the choke plate position, open the throttle plate with the engine cold, and allow the choke mechanism to operate. Then loosen the three screws that hold the choke housing retaining ring in place. Rotate the housing as required so that choke plate is held closed. Spring tension will vary according to climate. An extremely cold environment will require more spring preload so that the choke doesn't completely open before the engine warms up.

This small screw controls the position of the metering rods within their respective jets, and can consequently alter air/fuel ratio slightly. Rotating the screw counter-clockwise lowers the rods, providing more restriction and hence a leaner mixture. Clockwise rotation richens the mixture. This adjustment affects mixtures at part-throttle cruise, rather than full power, so it can have a significant impact on gas mileage.

CARTER AVS REBUILDING TIPS

The Carter AVS is often confused with the AFB, and for good reason—they are very much alike in both appearance and function. In fact, the primary sections of these carbs are almost identical. The greatest design difference lies in the secondaries. The AVS utilizes a spring-loaded "flap," rather than the counterweighted "throttle-plate" used in the AFB. The AVS also has no secondary booster venturis. Simple discharge tubes, like those used in the Thermo-Quad, serve as fuel nozzles. Since AVS carburetors were used primarily as original equipment, rather than aftermarket replacements, there is no standardization to the part numbers. Therefore, it is especially important to check the part number to insure that the proper rebuild kit is used.

DISASSEMBLY

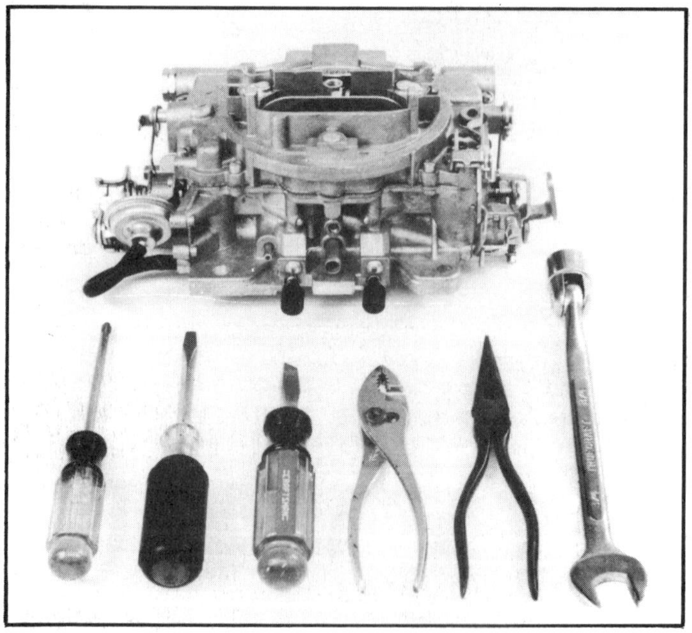

These are the only tools normally required to disassemble an AVS.

The carburetor part number can be found here, or on a tag which is held in place by one of the carb top screws. It's always a good idea to note the carb number for future reference.

Removal of the fuel line inlet fitting is a good place to start the disassembly process. Breaking the fitting loose is best done while the carb is still bolted to the intake manifold, as this "holds" the carb steady.

After the inlet fitting is removed, pull the fuel filter out. Its condition will give you good insight into the need for—or effectiveness of—external fuel filters.

In order to remove the air horn assembly, all the linkage will have to be disconnected. The secondary lockout link on the choke side of the carb may be dropped off by first removing the retaining clip.

Next the accelerator pump linkage is removed. A small clip holds the link rod in the actuating lever. Don't remove it before you note which hole the rod was originally inserted.

The lever retaining screw should be loosened, but it does not need to be removed until the carb top is lifted off and you wish to remove the accelerator pump piston.

Note the small spring used on this accelerator pump lever and install it as shown when the carb is reassembled—it controls the bowl-vent arm (arrow).

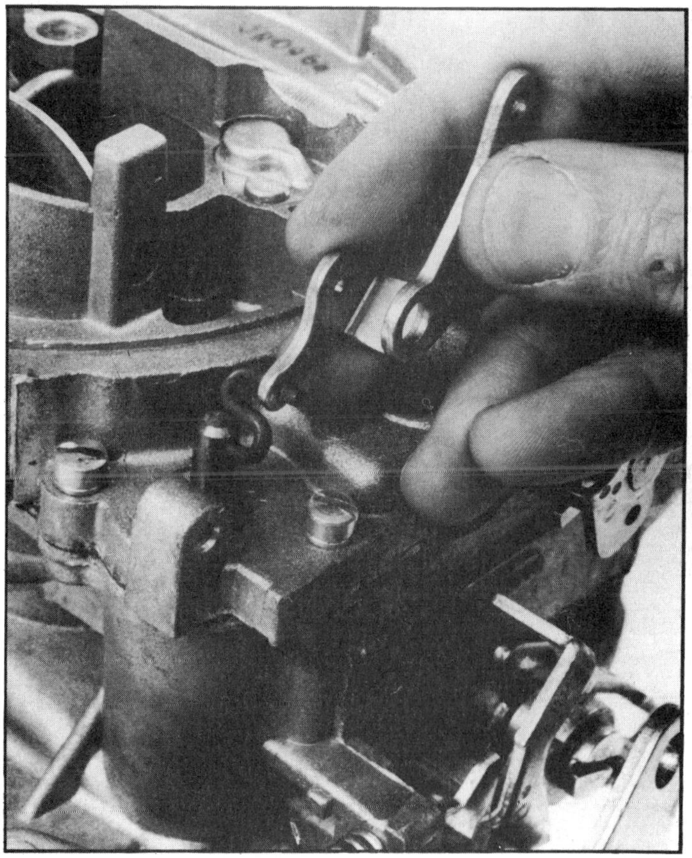

The accelerator pump "S" link will fit in one of two directions. It should always be installed so that it forms the letter "S," as opposed to "Z," when viewed from the front of the carburetor.

On some models, the choke pull-off linkage will have to be removed before the air horn can be persuaded to part company with the main body. The easiest method is to pull out the small clips on either end of the rod.

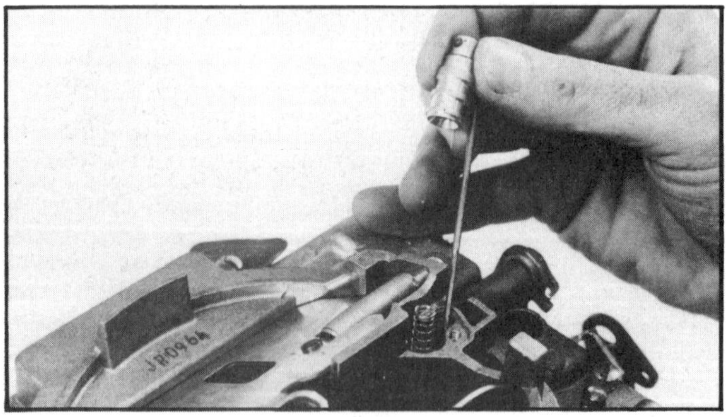

Lift the piston out and the metering rod will come with it, revealing the piston spring.

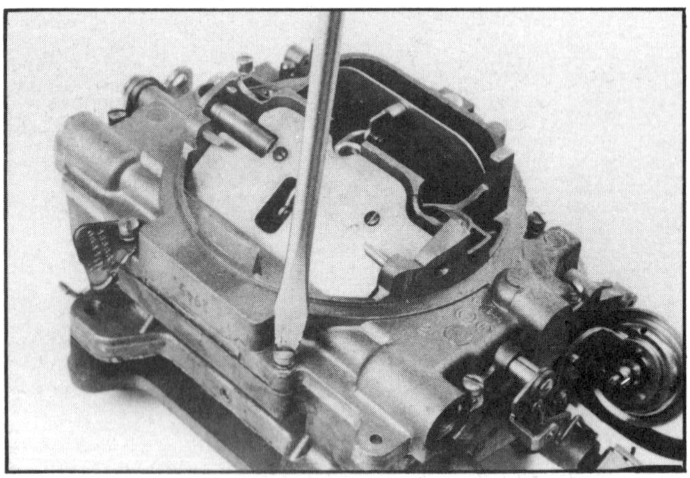

Now the screws holding the air horn to the main body can be removed. A total of eight screws are used, but all are not of the same length. Note the position of the various screws if you do not have a good memory for such trivia.

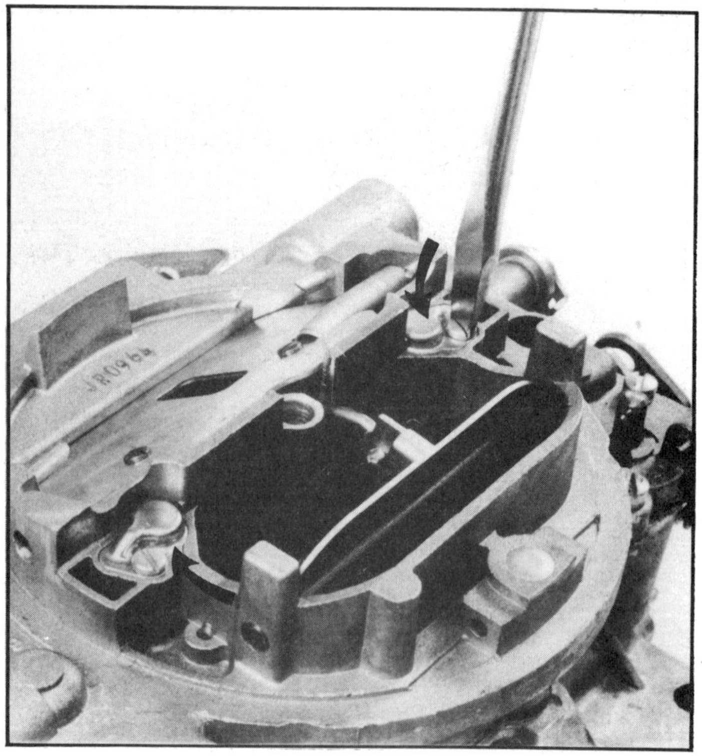

Each metering rod has its own piston and spring assembly located beneath these retaining plates (arrows). Remove the screws slowly while holding the plate as spring pressure can pop the piston out and send the plate soaring across your workbench.

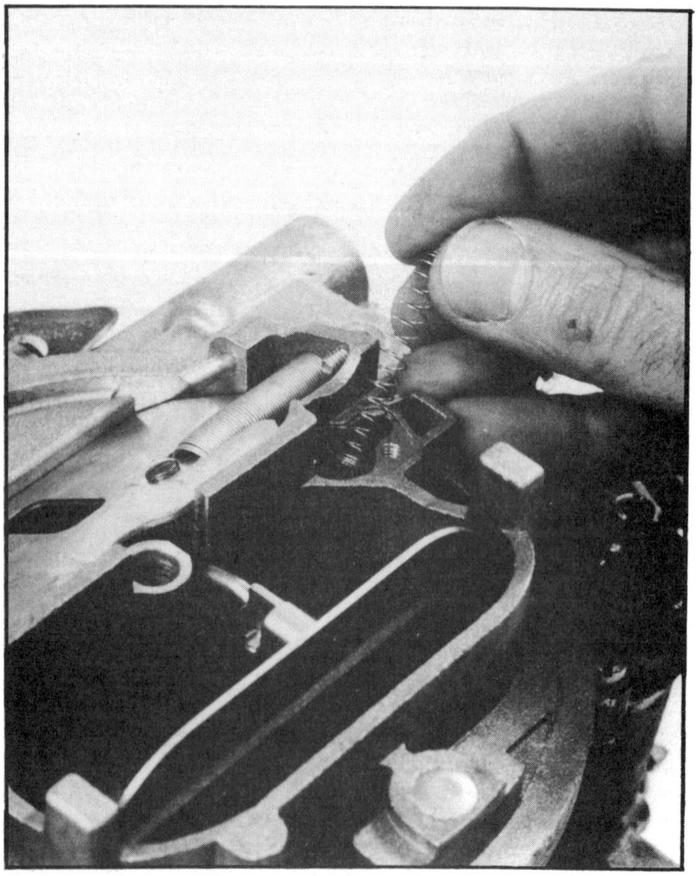

The piston spring tension controls metering rod operation, so be careful not to stretch or damage it.

With the screws removed lift the top off, being careful not to tilt it side to side, as the float-retaining pins can fall out. Remove the loosened accelerator pump arm and "S" clip. Then remove the accelerator pump piston (arrow) from the carburetor top.

Next remove the plastic caps that cover the idle mixture screws by giving them a pull (sometimes it helps to work them side to side—but do it gently, as too much pressure can bend or break the idle screws). The caps serve to limit adjustment for emissions purposes. While they are a nuisance, they're best retained—just in case the gendarmes decide to snoop around the engine compartment.

Lift the accelerator pump counterspring (arrow) out of the pump well.

Unscrew the idle-mixture adjusting screws. Inspect them and if they show signs of grooving on the taper, find replacements.

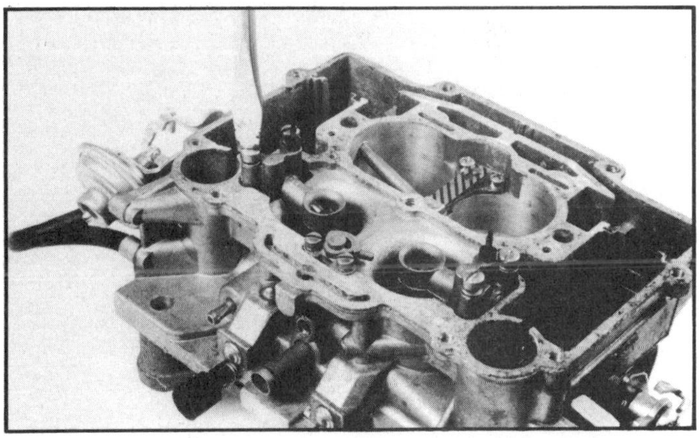

The primary booster venturis are held in place by two screws. Remove them and lift the booster out. You may need to wedge (gently) a screwdriver between the carburetor body and the booster to pop the gasket loose. When replacing the primary boosters, they should slip into place easily. If they don't, the left and right side pieces have been switched. One way to easily identify proper position is to note the brass air bleed tube on the top side. It should be toward the rear of the carburetor. Do not force the boosters into place or you may damage the brass tubes which extend from the bottom surface of the boosters.

Two screws hold the accelerator pump discharge nozzle in place. There's nothing tricky about removing the nozzle, except remembering to invert the carburetor and allowing the check needle to drop out of the fuel passage that lies below the nozzle.

The jets reside in the bottom of the float bowls. Use a wide-blade screwdriver to remove them, or you may damage the slot.

This is the hot-idle compensator. It allows additional air to bleed into the intake manifold during high-temperature operation. Remove it by loosening the two Phillips retaining screws.

Fuel is admitted to the accelerator pump chamber through this inlet check valve. It looks similar to a jet, but the internal valve mechanism is obvious once it's removed.

Two screws retain the choke pull-off diaphragm. This particular carb uses a divorced choke, so there is comparatively little hardware to remove. (See AFB section for comparisons and information pertaining to AVS carbs with integral chokes.)

Start stripping the air horn assembly by removing the float pins and floats.

Then lift out the inlet check needle.

Unscrew the inlet seat with a wide-blade screwdriver.

Reverse this procedure to reassemble the carb. And remember, don't force anything. If a component doesn't slip into place, something is wrong. Fix it before proceeding or you'll regret your haste later.

ADJUSTMENTS AFTER REBUILDING

Set the floats by inverting the air horn assembly and measuring distance between gasket and float. Check the factory manual for specific recommendations. If the factory specifications are unavailable, start with 9/32-inch. To adjust, bend float lever with a pair of pliers. DO NOT push float down using needle-and-seat as a fulcrum as this may damage the tip of the needle.

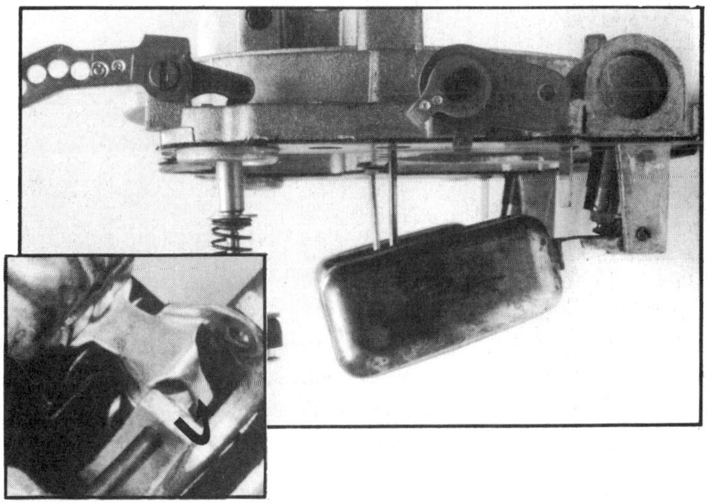

Float drop is controlled by the tab that rests against the inlet seat. Set according to individual carb recommendations as listed in factory manual. This setting is not as critical as float height, however, insufficient drop can restrict inlet fuel flow.

To adjust throttle plate position at wide open throttle, invert the carburetor and push throttle lever against its opening stop. Both sets of throttle plates should be at 90 degrees from carb base.

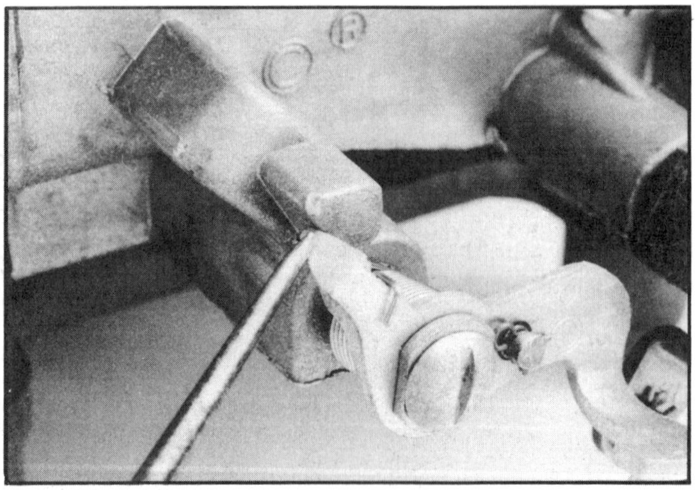

To adjust the secondary wide-open-throttle position, bend the link rod or the stop on the throttle lever.

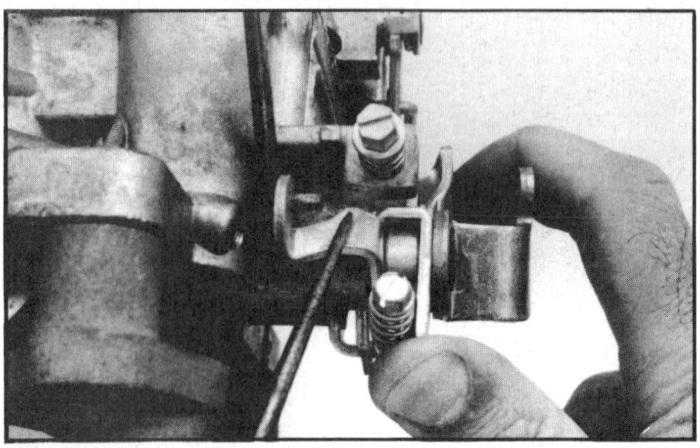

The primary throttle stop is this tang which bottoms against the carb body. Although adjustment is usually not necessary, it may be bent or filed if primaries do not open fully.

After rebuilding and reassembly, you can establish an initial idle mixture setting by rotating adjusting screws clockwise until they bottom (gently), and then backing them out 1-1/2 to 2 turns. After the carb has been installed on the engine, optimize setting by using a vacuum gauge or tachometer. Adjust for highest reading then replace the plastic caps.

Once the mixture setting has been established and the carburetor is reinstalled—with the engine running—adjust idle speed setting as shown above.

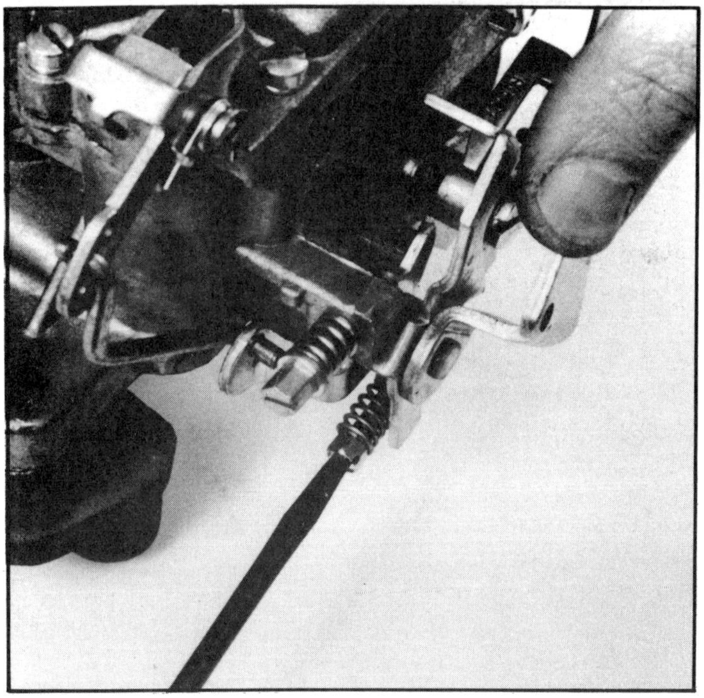

In order to adjust fast idle speed (fast idle occurs when the choke is operating—i.e., partly closed—during cold engine operation), rotate the screw shown above. With a cold engine make sure the hot-idle cam can move freely through its cycle (doesn't stick or bind).

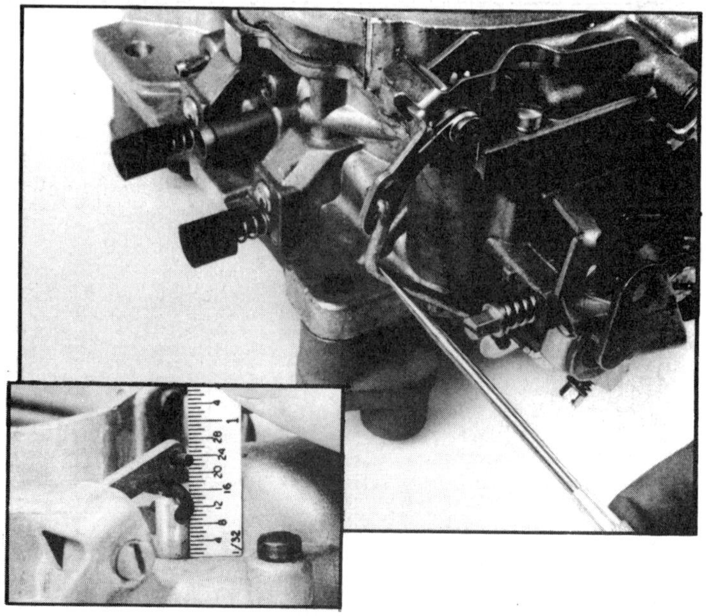

Individual applications may require other settings, but for most performance installations, the pump stem should be set to extend 13/32-inch above the top surface of the air horn with the idle-speed screw backed out and the throttle plates completely closed. Adjust the pump stem height by bending the link rod at the indicated point. Don't forget to reset idle speed.

To quicken the secondary opening rate, adjust the air valve spring tension using a wide-blade screwdriver to rotate the spring-load tensioner and a normal screwdriver (illustrated above) to loosen/tighten the lock screw. Rotating the spring tensioner counter-clockwise will increase tension, which slows the opening rate. Rotating the tensioner clockwise has the opposite effect. Holding the spring tensioner with one screwdriver, loosen the lock screw with another, then rotate the tensioner as necessary. Tighten lock screw before releasing tension on spring.

The choke pull-off is adjusted by bending this rod as shown. If improperly adjusted, the engine will run too rich when cold and the secondaries may not open when normal operating temperature is reached.

AFB REBUILDING TIPS

Rebuilding the Carter AFB is a straightforward process. The design is uncomplicated and assembly and disassembly can be accomplished—on the whole—without special tools. The AFB has only two major castings, the airhorn and main-body/throttle-plate assemblies. All four booster venturis are removable—permitting easy and thorough cleaning—but care must be taken to avoid damaging the brass tubes that protrude from the venturi bodies.

DISASSEMBLY

Removal of the fuel line fitting is a good place to start the disassembly process. Breaking the fitting loose is best done while the carb is still bolted to the intake manifold, since considerable torque may be required.

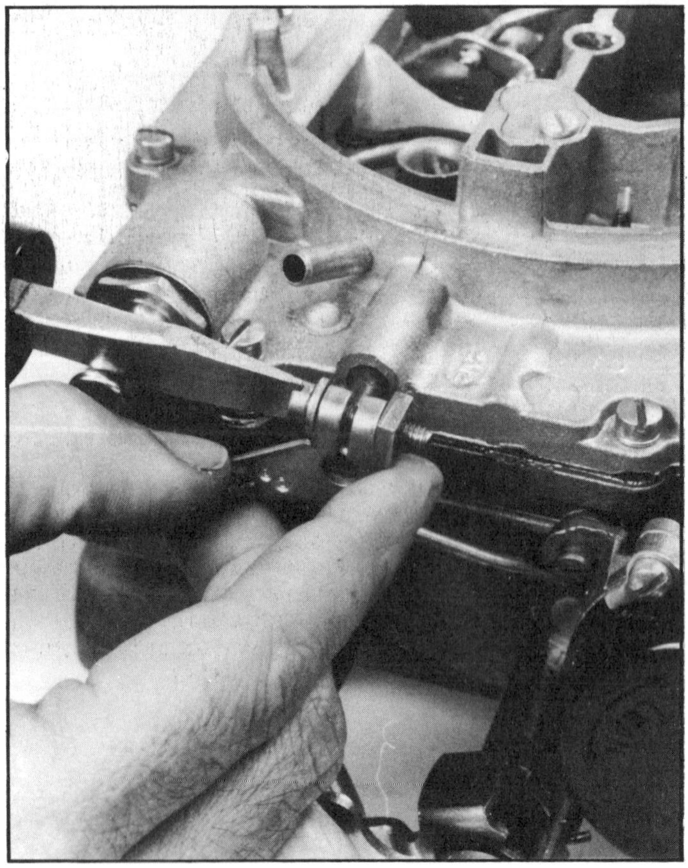

In order to remove the air horn assembly, the choke linkage must be disconnected. This includes the secondary lockout-link—on the choke side of the carb—which can be "dropped off" by loosening the lever retaining screw, removing the lever, disconnecting the link, and then reinstalling the lever on the choke rod (to prevent losing the lever and to aid in reestablishing proper assembly).

This is the other end of the secondary lockout-linkage. Remove this clip too.

Now rotate the secondary lockout-link rod so that the key on the lower end lines up with the slot in the lever. Remove the rod, but note into which hole it originally fit (arrow).

Next, the accelerator pump linkage is removed. A small clip holds the link rod in the actuating lever; don't remove it before noting into which hole the rod was originally inserted. The clip is most easily removed with needle-nose pliers, however, the spring tension in the clip is small, and almost any small pointed tool will work (be sure to wear eye protection during rebuilding to prevent getting a "close look" at flying clips and splashes of carburetor cleaner).

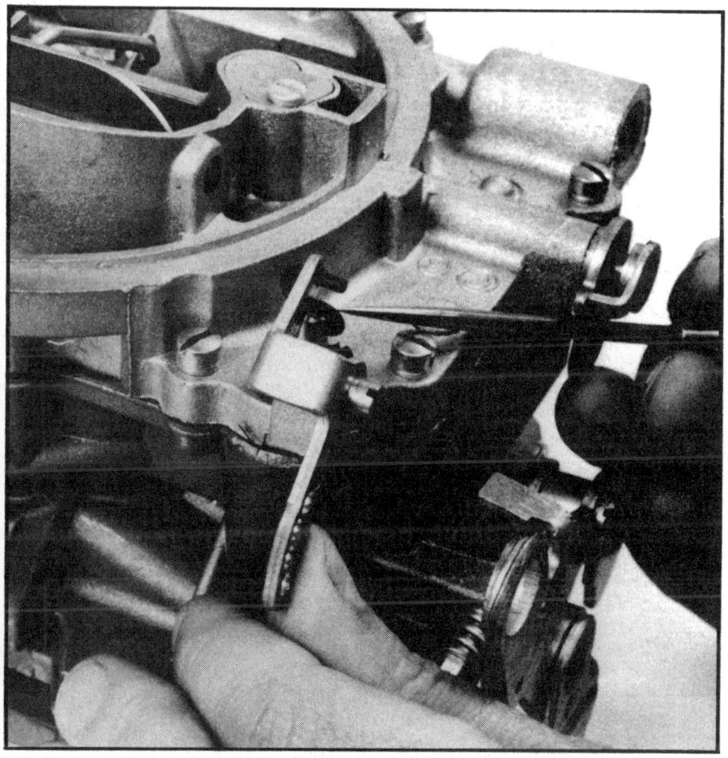

The lever retaining screw (left) should be loosened, but it does not need to be removed until the carb top is lifted off. Lever removal (along with the "S-shaped" connecting link—right) will "disconnect" accelerator pump piston from the air-horn assembly. The "S" link is correctly reinstalled in only one of two directions; it should form the letter "S," as opposed to "Z," when viewed from the front of the carburetor.

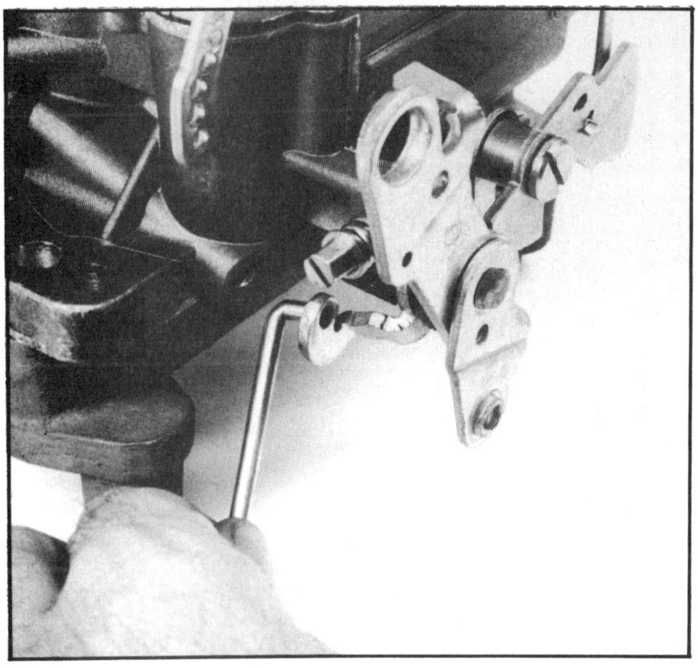

Once unhooked from the pump lever, the link rod may be rotated downward and removed from the throttle linkage.

Each metering rod piston is spring loaded against these retaining plates. Remove the screws slowly while holding the plate as the piston can fly out and send the retaining plate on a difficult-to-retrace journey.

Lift the piston out and the metering rod will come with it, revealing the piston spring.

The piston spring tension controls metering rod operation, so be careful not to stretch it out of shape.

Now the screws holding the air horn to the main body can be removed. A total of ten screws are used, but all are not of the same length. Note the screw positions for reassembly, if you don't have a good memory for such trivia.

With the screws removed, lift the top off being careful not to tilt it side to side; the float retaining pins (arrows) can fall out.

Lift the accelerator pump counter spring out.

Next, remove the idle-mixture adjusting screws. Inspect them and if the tapered ends have been badly grooved, find replacements.

The accelerator pump discharge nozzle is held in place by two screws. There's nothing tricky about removing the nozzle.

However, you must remember to invert the carburetor and allow the check needle to drop out. Forgetting this step can lead to worries about where it came from when it suddenly appears on the work bench.

The primary booster venturis are held in place by two screws. You may need to wedge (gently) a screwdriver between the carburetor body and the booster to pop it loose. Do not use excessive force! When replacing the primary boosters, they should slip into place easily. If they don't, the left and right side pieces have been switched. Improper installation will certainly damage the brass tubes which extend from the bottom surface. One way to easily identify proper location is to note the rearward position of the brass air bleed tube on the top side.

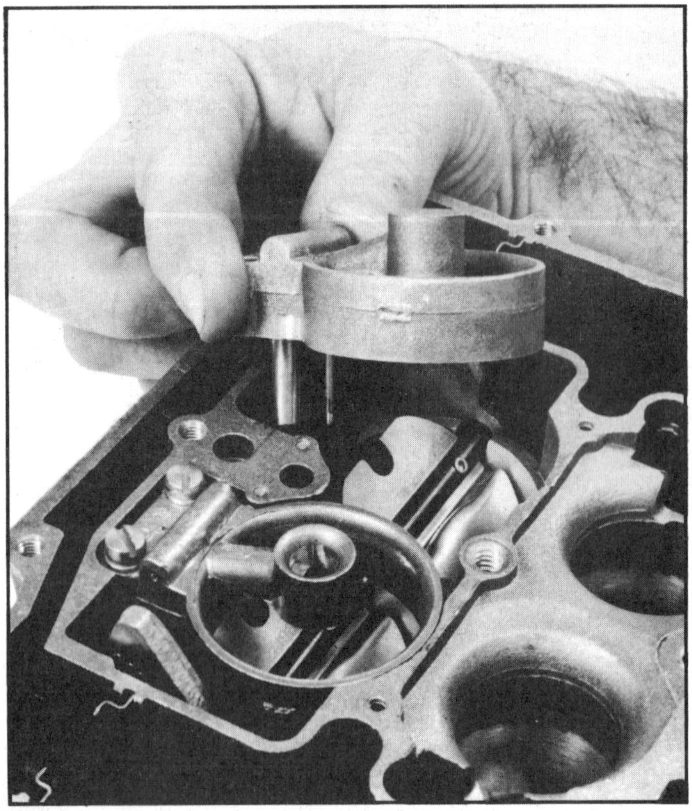

The same procedure may be used on the secondary boosters. You need not note their position as they cannot be improperly installed.

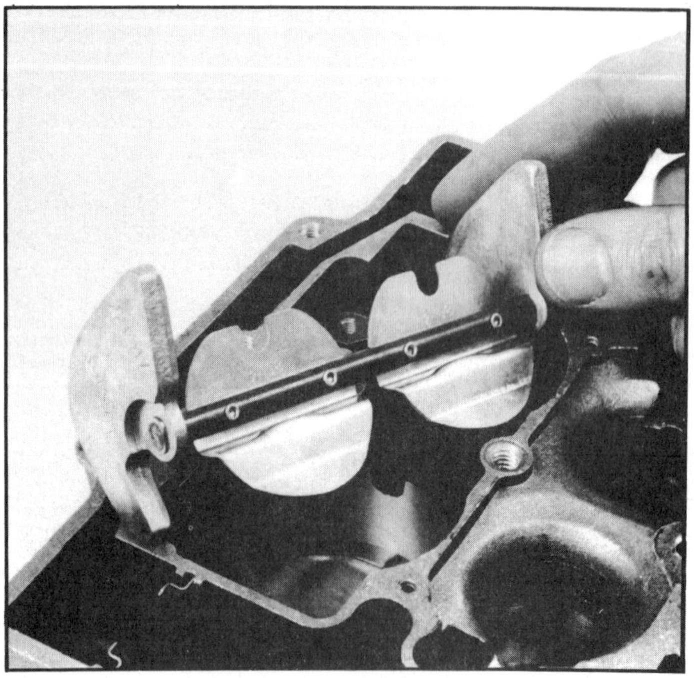

With the secondary boosters out of the way, the air valve may be lifted out.

The four main-metering jets reside in the bottom of the float bowl (two are indicated by arrows). Use a wide-blade screwdriver to remove them.

Three screws retain the choke housing. Before loosening, note the relationship of the black plastic housing to the cast metal piece so that the choke setting may be easily reestablished.

The backing plate and gasket may be lifted from the choke once the plastic housing is removed.

Three screws hold the choke mechanism to the main body. Two are beneath the plastic cover and one is external. Obviously, all three must be removed.

Manifold vacuum is applied to the choke housing to activate the choke pull-off and to draw in hot air through the hot-air choke tube. A vacuum seal is maintained by this gasket. It should be replaced when the carb is reassembled, preventing leaks that will impair choke operation.

The pull-off plunger may be removed by loosening this screw.

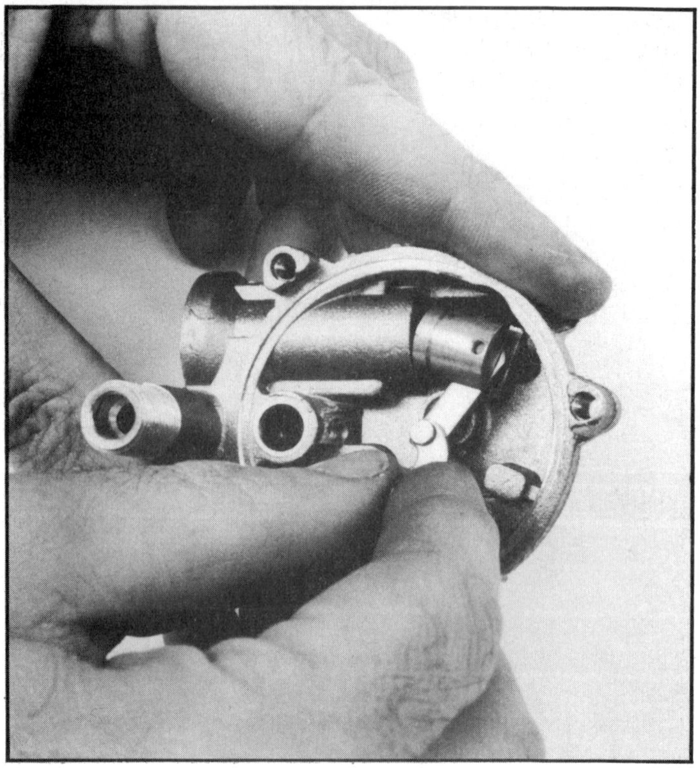

Once the screw is loose, rotate the lever rearwards. The plunger will often become gummed up or frozen in its bore during normal operation. It should be thoroughly cleaned to ensure proper choke operation.

The accelerator pump side of the AFB appears to have three jets. The one located near the pump chamber is actually an inlet check assembly. It too should be removed prior to dipping the carb in "the vat."

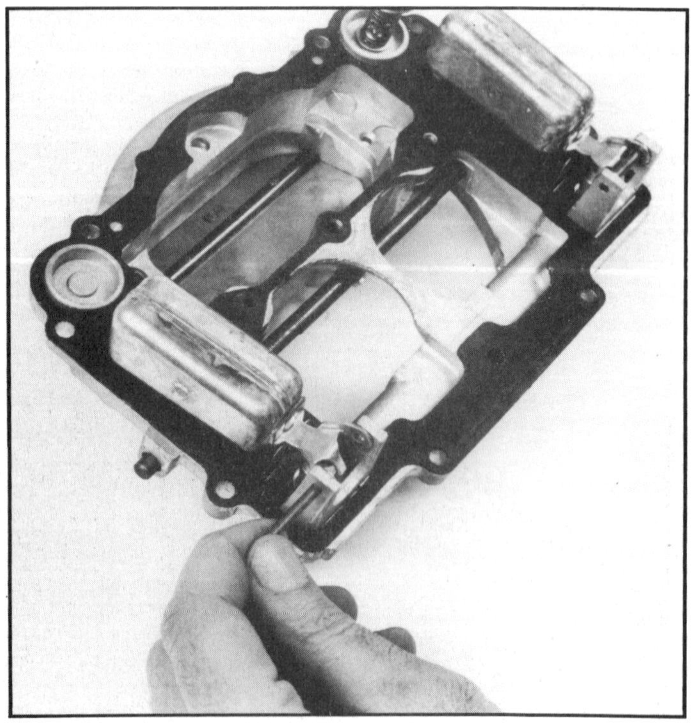

Start stripping the air horn assembly by removing the float pins and floats.

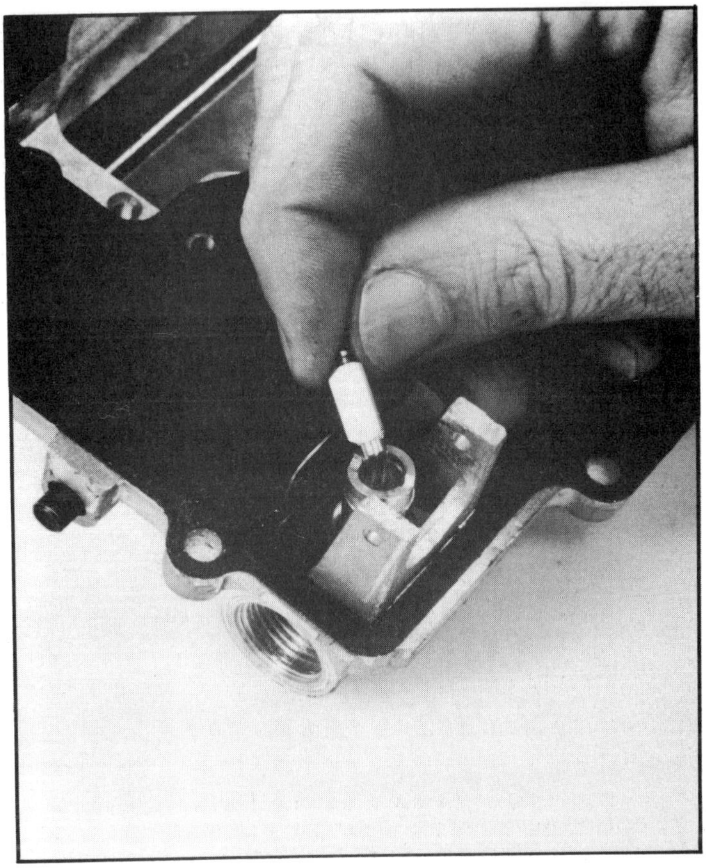

Then lift out the inlet check needles.

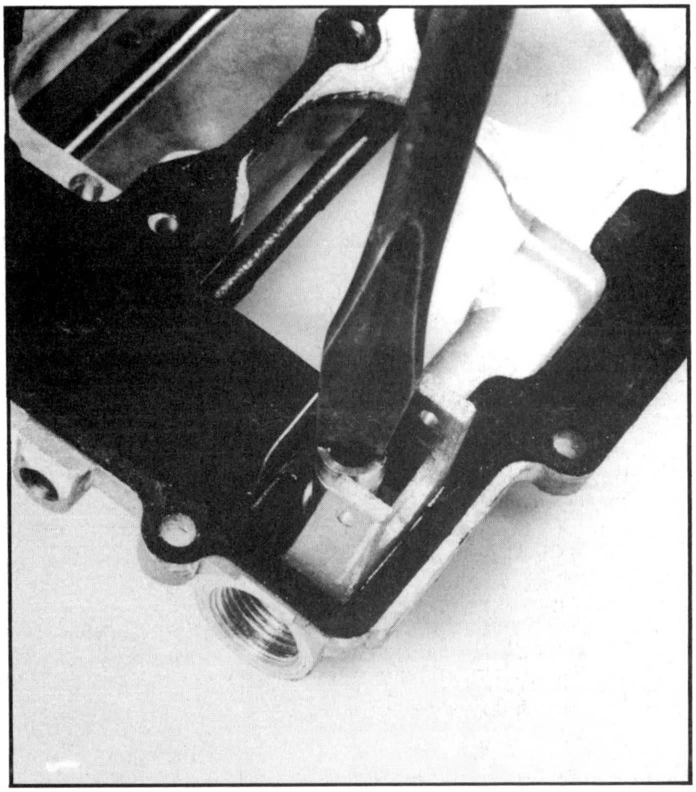

Unscrew the inlet seats with a wide-blade screwdriver.

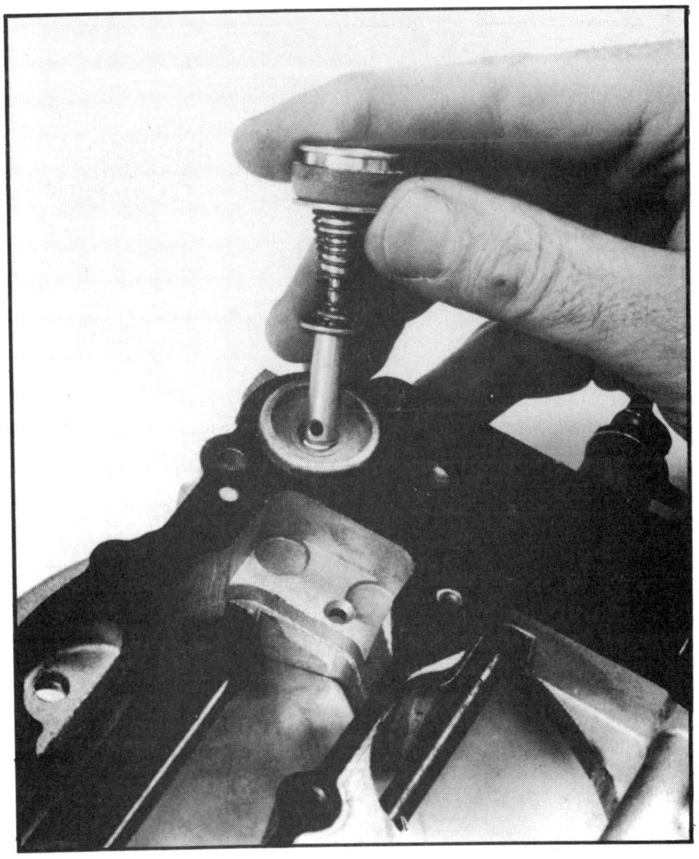

Once the "S" link is removed, the accelerator pump plunger will slide out.

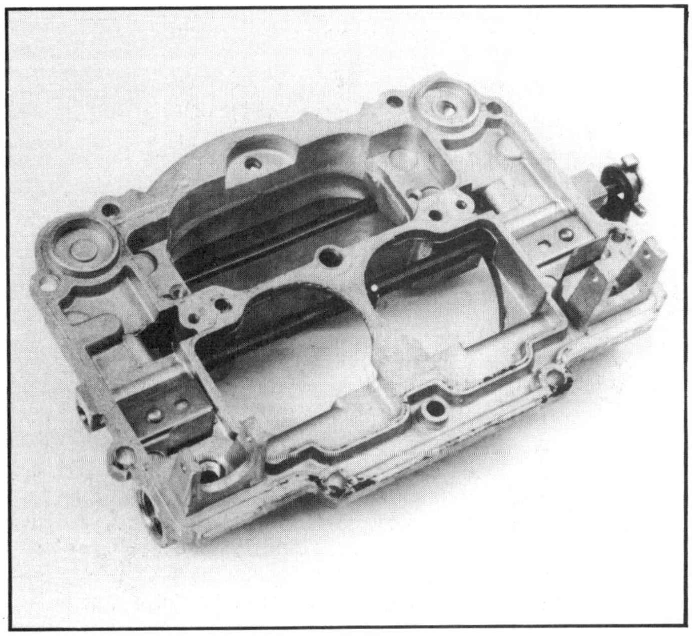

After removing all traces of the air-horn gasket, the completely stripped air horn may now be soaked in carburetor cleaner.

Reverse the disassembly procedure to reassemble the carb. And remember, don't force anything. If a component doesn't slip into place, something is wrong. Correct the orientation before proceeding or you'll regret your haste later.

ADJUSTMENTS

Set the floats by inverting the air horn assembly and measuring distance between gasket and float. With 9000 Series AFB's this dimension should be 9/32-inch. To adjust, bend float lever with a pair of pliers. DO NOT push float down using needle-and-seat as a fulcrum as this may damage the needle tip.

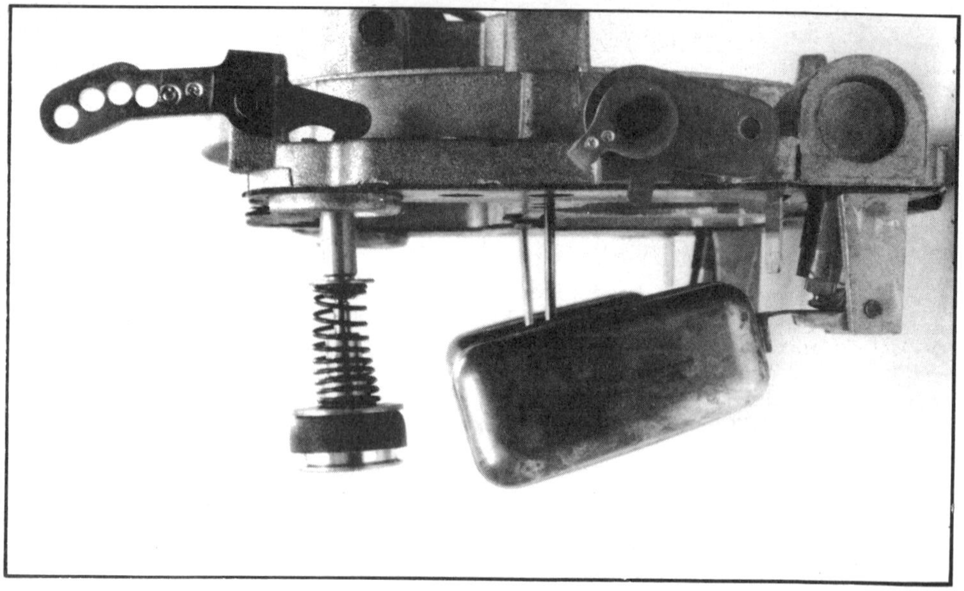

Float drop is controlled by the tab that rests against the inlet seat. On 9000 Series carbs, drop should be set to 15/16-inch.

To adjust throttle plate position at wide open throttle, invert the carburetor and push throttle lever against its stop. Both sets of throttle plates should be at 90 degrees from carb base.

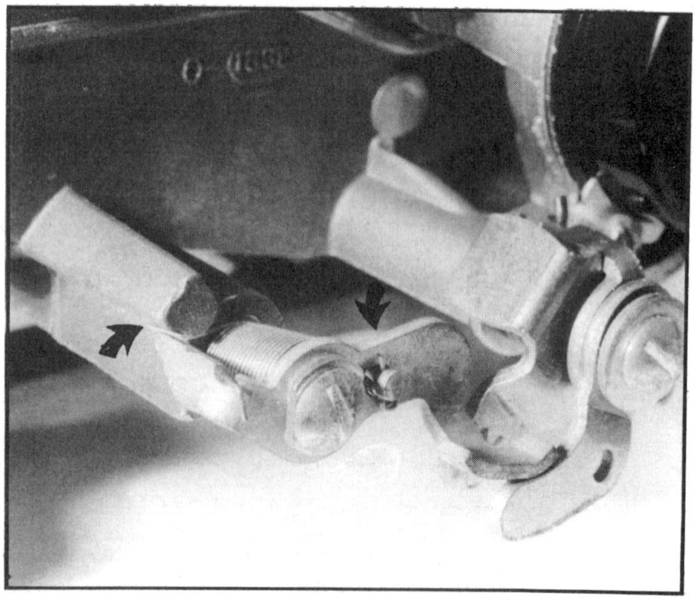

To adjust secondaries, bend the link rod or the stop on the throttle lever.

The primary throttle stop is this tang which bottoms against the carb body. Although usually not necessary, it may be bent or filed if primaries do not open fully.

Establish an initial idle-mixture setting by rotating adjusting screws clockwise until they bottom (gently) and then backing them out 1-1/2 to 2 turns. After the carb has been reinstalled on the engine, optimize setting by using a vacuum gauge or tachometer; normally the optimum idle-mixture setting occurs at the highest obtainable vacuum/rpm reading.

Individual applications may require other settings, but for most performance installations, the pump stem should be set to extend 13/32-inch above the top surface of the air horn; the idle-speed screw should be backed out completely ensuring the throttle plates are completely closed. Adjust by bending the link rod (at arrow). Don't forget to reset idle speed.

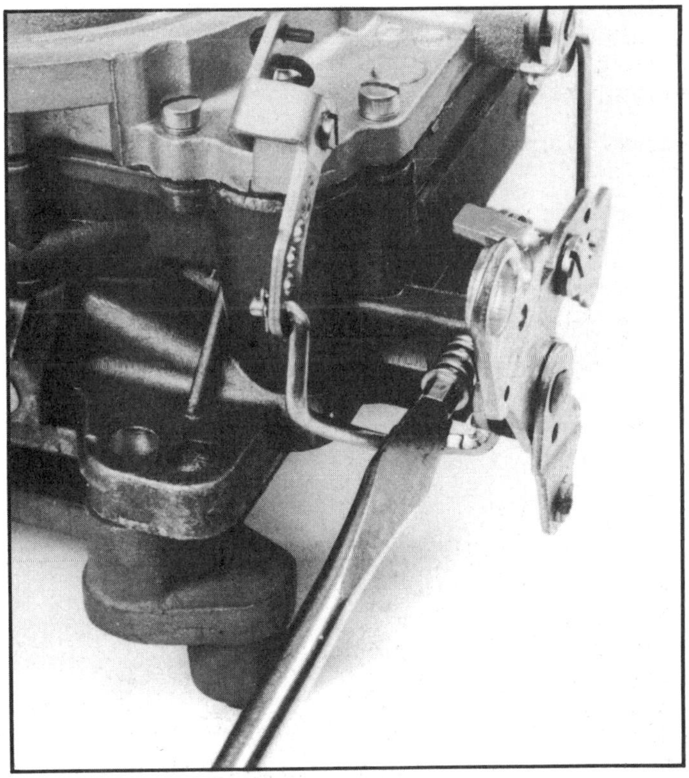

Once the mixture setting has been established, adjust idle speed setting as required.

WCFB REBUILDING TIPS

The Carter WCFB was the first four-barrel carburetor used by American auto manufacturers. Production ceased in the 1960s, but there is still considerable interest in this model because it was commonly used on vehicles that have become popular with collectors and restorers. Most notably, a pair of WCFB carbs were used on the 270hp, 283 cubic-inch engine used in Corvettes and Chevrolets during the late 1950s and early 1960s.

It is a relatively simple carburetor to rebuild and overhaul kits are still readily available. However, reviving a WCFB that has sustained damage to any of its castings may be difficult as replacement pieces are available only in wrecking yards, or through a coterie of rebuilders who have accumulated pieces over the years.

Before tearing into a WCFB, familiarize yourself with its external linkages so that they may be easily reinstalled in their proper positions. There were several variations of the WCFB, so your carburetor may not look exactly like this one.

The first step is to disconnect the choke and accelerator pump linkage prior to removing the air horn assembly screws. Loosening the screw on the choke link lever will allow it to slip off. The accelerator pump rod may be pulled from the hole in the transfer lever once the retaining clip (arrow) has been removed.

Before popping the top of the carb, it's good practice to remove the choke housing. Note the position of the plastic housing (relative to the metal portion of the choke housing), then loosen the three retaining screws.

The plastic housing can now be removed, exposing the interior of the choke mechanism and the choke pull-off piston. Three screws attach this housing to the air horn assembly; one shown at right (arrow) and two others that are inside the housing (indicated by the position of the screwdrivers).

Continuing to work with the air horn assembly, remove the metering rod dust cover and the two set screws that hold it in position. This will reveal the metering rod and accelerator pump actuating mechanism. Both circuits are controlled by the same rod.

However, before the metering rod hanger can be removed, the power piston must be positioned as shown and unhooked (the air horn assembly is removed and inverted in this picture).

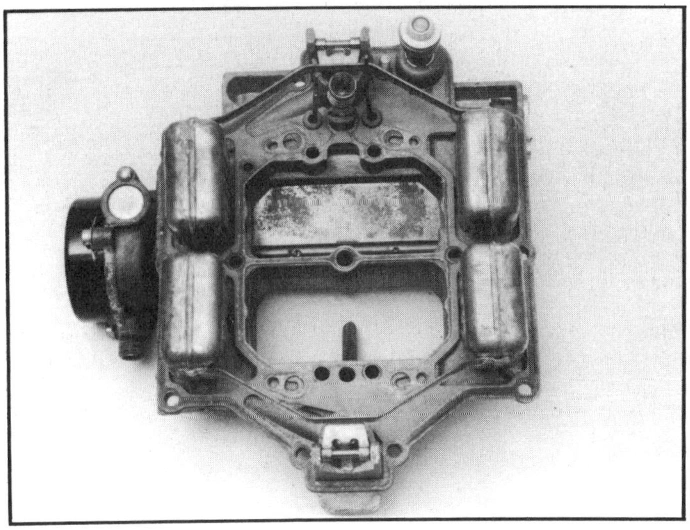

Proceeding to the underside of the air horn, note that two separate float assemblies are utilized. While these are frequently interchanged with few consequences, it is advisable to identify primary and secondary floats in order to minimize the amount of time resetting float drop and level.

In order to remove the metering rods and their connecting bracket (called a "hanger"), the screw that holds the limit arm (indicated by screwdriver) to the operating shaft must be loosened. Once this is done, the arm can be rotated back so that the hanger may be pulled upward. The same procedure should be followed to remove the accelerator pump piston.

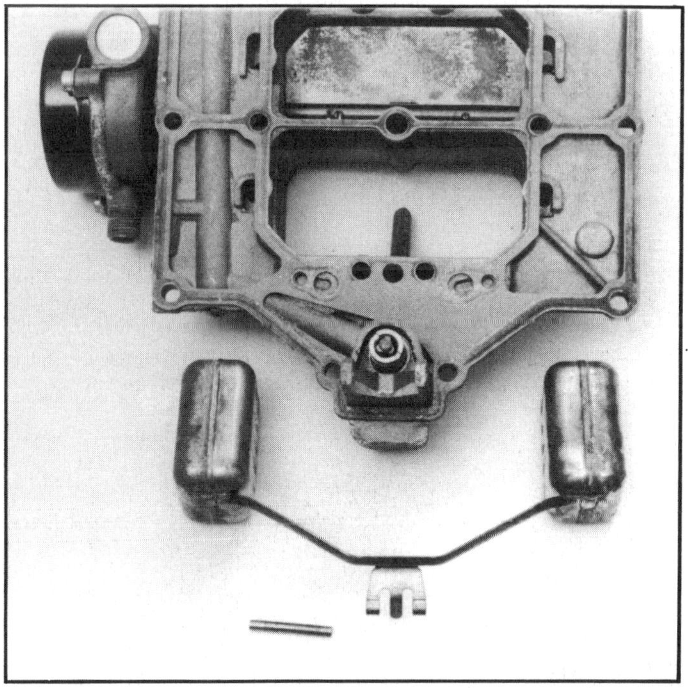

The float assembly may be removed by simply sliding the pivot pin out of mounting. Then the needle may be pulled out and the seat assembly removed with a wide-blade screwdriver.

Stripping the main body may begin by removing the four main metering jets and accelerator pump inlet check valve in the floor of the float chambers (arrows).

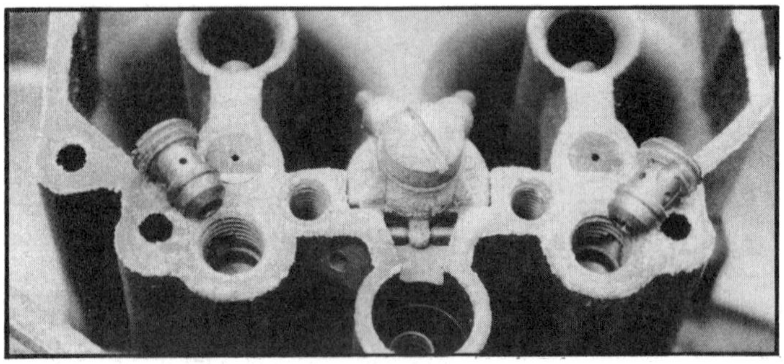

Some models have removable low-speed jets on the primary side. These should be taken out before sending the carburetor to the "soup."

Once the main-body screws are removed, the throttle plate assembly can be detached from the main body casting.

The accelerator pump spring (right) can now be removed from the pump well. Next remove the metering-rod-piston spring (left) from the piston chamber (arrow). Finally, remove the pump nozzle and check valve from their location between the primary booster venturis.

The throttle body is attached to the main body by six screws.

Before discarding the original throttle-body/main-body gasket, check it against the one (or several) supplied in the overhaul kit to ensure that all holes are in the proper position; discard any incorrect gaskets to prevent accidental installation.

Once the mixture screws have been set, idle speed may be adjusted as desired by turning the screw shown by arrow "1." The adjusting screw—arrow "2"—controls fast-idle speed when the choke is engaged. This setting should be made with a cold running engine, and the choke in its normal "cold-start" position.

The idle-mixture screws should be checked for ridges or other damage that may interfere with smooth adjustment. When rotated clockwise, these screws lean the idle mixture. When resetting them, rotate clockwise until they bottom (gently), then turn them counter-clockwise 1-1/2 turns. Once the carburetor is installed on the engine, the screws should be reset to provide the highest manifold vacuum reading and/or idle speed.

With wear and tear over the years, the secondary-actuation throttle shaft arms may have become maladjusted. Move the throttle lever to wide open throttle position and then check the linkage to ensure that the arms are contacting the stops on the throttle body (one is visible in this photo—arrow) Before making any adjustments, check the position of throttle plates. If the primary and secondary throttles are not open equally, bend the link rod behind the arms. Once uniform opening is accomplished, bend the arm tabs so that they contact the throttle-plate stops at the wide-open position.

THERMO-QUAD EXPLODED VIEW

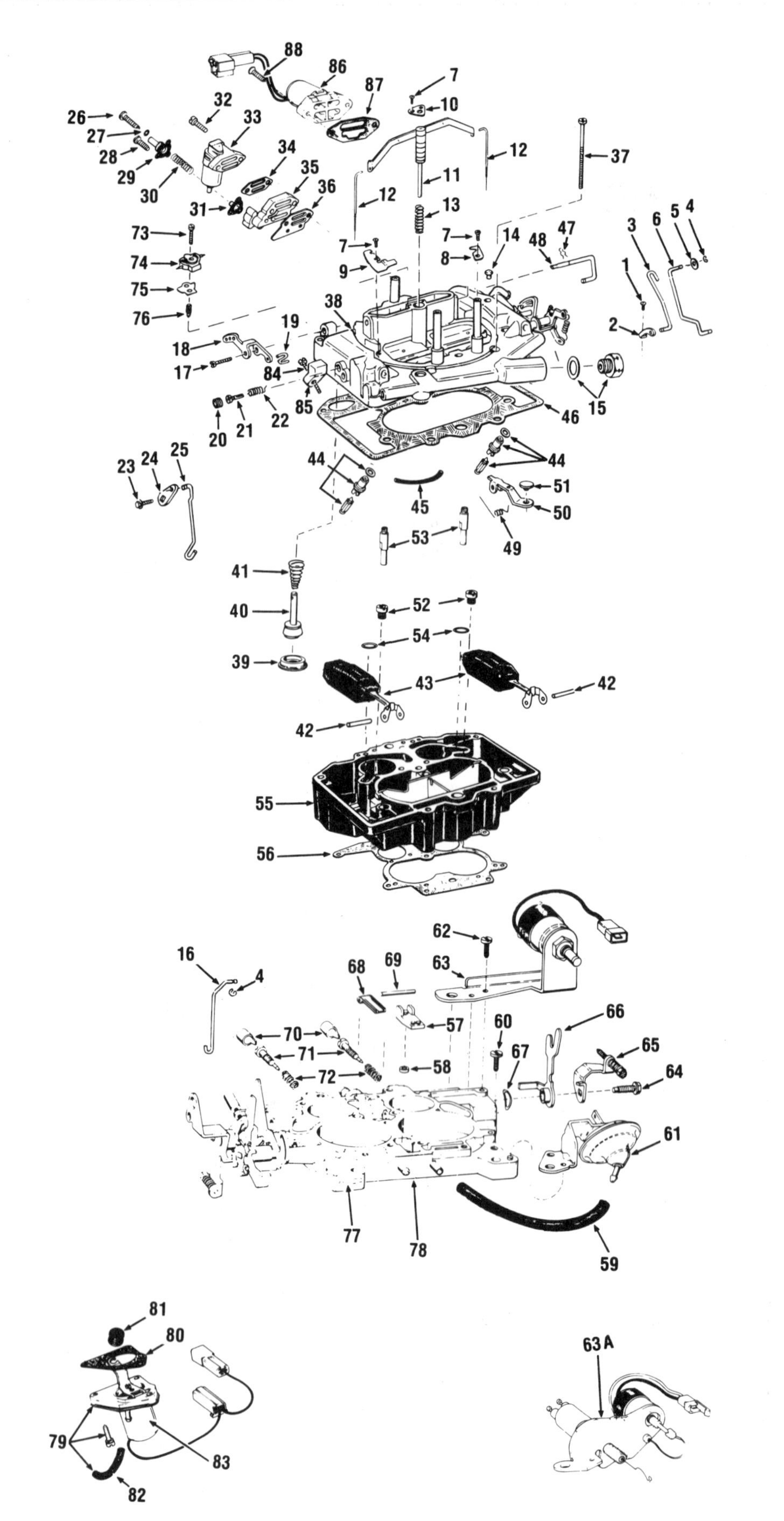

PARTS LISTING

1. Choke lever screw
2. Choke lever
3. Choke connector rod
4. Choke pull-off rod retainer (2)
5. Choke pull-off connector washer
6. Choke pull-off connector rod
7. Step-up piston and metering rod cover plate screw
8. Metering rod cover plate (choke side)
9. Metering rod cover plate (pump side)
10. Step-up piston cover plate
11. Step-up piston assembly
12. Metering rod (2)
13. Step-up piston spring
14. Bowl vent adjustment plug
15. Fuel inlet fitting and gasket
16. Throttle connector rod
17. Pump arm screw
18. Pump arm
19. Pump connector "S" link
20. Air valve lock plug
21. Air valve adjustment plug
22. Air valve spring
23. Countershaft scres
24. Countershaft lever
25. Fast idle connector rod
26. Idle enrichment screw
27. Idle enrichment cover screw washer
28. Idle enrichment cover screw (2)
29. Idle enrichment diaphragm cover
30. Idle enrichment diaphragm spring
31. Idle enrichment diaphragm
32. Altitude compensator screw (2)
33. Altitude compensator
34. Altitude compensator gasket
35. Idle enrichment casting
36. Idle enrichment casting gasket
37. Bowl cover screw (10)
38. Bowl cover assembly
39. Pump intake check assembly
40. Plunger assembly
41. Plunger spring
42. Float pin (2)
43. Float (2)
44. Needle, seat and gasket
45. Pump passage tube
46. Bowl cover gasket
47. Bowl vent pin retainer
48. Bowl vent lever (upper)
49. Bowl vent lever spring
50. Bowl vent arm
51. Bowl vent grommet
52. Primary metering jet (2)
53. Secondary metering jet (2)
54. Quad "X" rings-1972 and later (2)
55. Fuel bowl
56. Flange gasket
57. Idle compensator valve
58. Idle compensator gasket
59. Choke pull-off hose
60. Choke pull-off bracket screw
61. Choke pull-off and bracket
62. Bracket screw (if used)
63. Solenoid and bracket

63A. Transducer and bracket

64. Bowl vent lever operating screw
65. Bowl vent lever (lower)
66. Bowl vent fork lever
67. Throttle shaft washer
68. Step-up piston lifter
69. Step-up piston lifter rod
70. Idle limiter cap (2)
71. Idle mixture screw (2)
72. Idle mixture screw spring (2)
73. Pump housing screw
74. Pump jet housing
75. Pump housing gasket
76. Pump discharge check needle
77. Carburetor identification number
78. Flange assembly
79. Bowl vent valve screw (3)
80. Bowl vent valve gasket
81. Bowl vent valve grommet
82. Bowl vent valve hose
83. Bowl vent solenoid
84. Low idle speed screw
85. Low idle speed screw spring
86. Pulse solenoid
87. Pulse solenoid mounting gasket
88. Pulse solenoid mounting screws (2)

PARTS LISTING	AVS EXPLODED VIEW

1. Step up piston cover screw (2)
2. Step up piston cover plate (2)
3. Step up piston and rod assembly (2)
4. Vacuum piston spring (2)
5. Fuel inlet fitting
6. Fuel inlet fitting gasket
7. Fuel inlet fitting screen
8. Pin spring (small)
9. Retainer

9A. Pin spring (large)

10. Washer
11. Diaphragm link
12. Throttle connector rod
13. Fast idle rod
14. Pump arm screw
15. Vent arm spring
16. Vent arm
17. Vent valve
18. Pump arm
19. Pump link
20. Air horn screw (8)
21. Solenoid bracket screw (2)
22. Solenoid and bracket
23. Air horn assembly
24. Float lever pin (2)
25. Float (2)
26. Needle and seat assembly (2)
27. Air horn gasket
28. Pump plunger
29. Pump plunger spring
30. Pump intake check
31. Primary venturi screw (4)
32. Primary venturi vent seal (2)
33. Primary venturi (2)
34. Primary venturi gasket (2)
35. Pump jet screw (2)
36. Pump jet housing
37. Pump jet gasket
38. Pump discharge check needle
39. Primary jet (large orifice) (2)
40. Secondary jet (small orifice) (2)
41. Compensator valve screw (2)
42. Compensator valve
43. Compensator valve gasket
44. Idle mixture screw and spring (1968, early 1969)
45. Idle mixture screw and spring (2)
46. Idle limiter cap (2)
47. Diaphragm and bracket screw (2)
48. Diaphragm and bracket
49. Diaphragm suction hose
50. Throttle speed screw

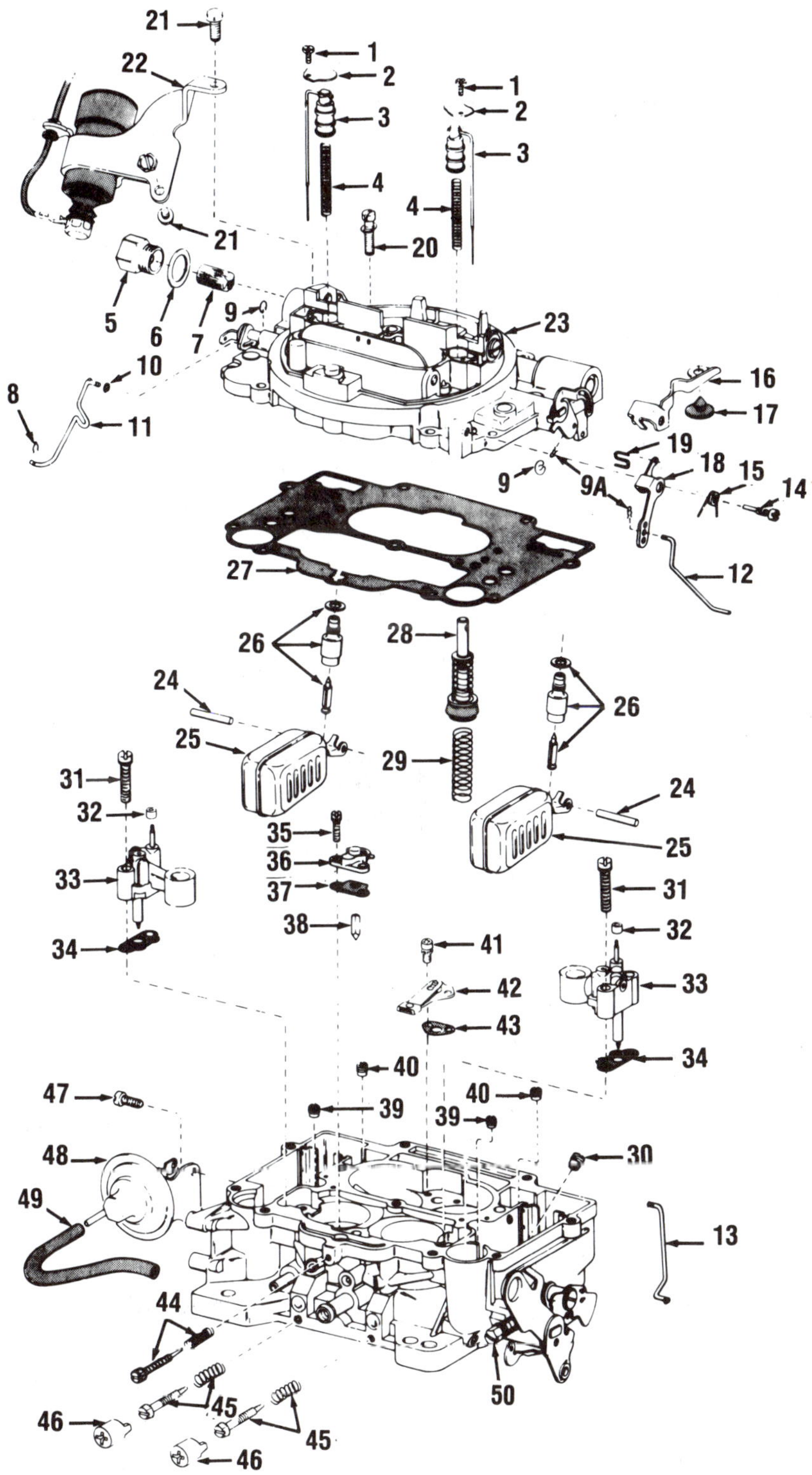

AFB EXPLODED VIEW

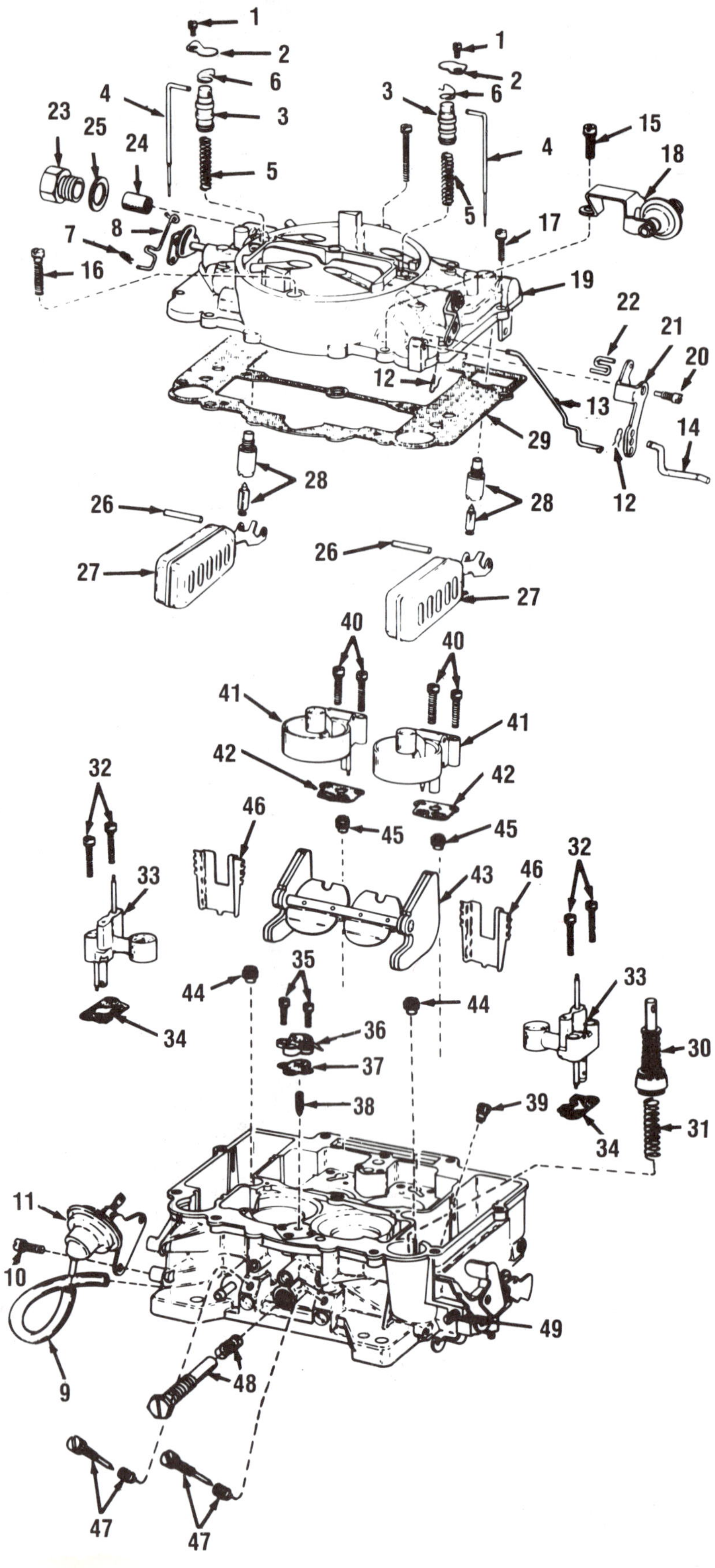

PARTS LISTING

1. Step up piston cover plate screw (2)
2. Step up piston cover plate (2)
3. Step up piston (2)
4. Step up rods (2)
5. Step up piston springs (2)
6. Step up rod retainer spring (2)
7. Pin spring (small)
8. Choke diaphragm connector rod
9. Hose
10. Choke diaphragm bracket attaching screws (2)
11. Choke diaphragm and bracket assembly
12. Pin spring (4)
13. Choke connector rod
14. Throttle connector rod
15. Bowl cover attaching screw
16. Bowl cover attaching screw
17. Bowl cover attaching screw (8)
18. Dashpot and bracket assembly
19. Air horn assembly
20. Pump arm screw
21. Pump arm
22. Pump connector link
23. Fuel inlet fitting
24. Strainer
25. Fuel inlet fitting gasket
26. Float lever pin (2)
27. Float and lever assembly (2)
28. Needle and seat assembly (2)
29. Air horn gasket
30. Pump plunger rod and spring assembly
31. Pump plunger spring
32. Primary venturi cluster screws (4)
33. Primary venturi clusters (2)
34. Primary venturi cluster gasket (2)
35. Pump jet housing screw (2)
36. Pump jet housing
37. Pump jet housing gasket
38. Pump discharge check needle
39. Pump intake check
40. Secondary venturi cluster screws (4)
41. Secondary venturi cluster (2)
42. Secondary venturi cluster gasket (2)
43. Auxiliary valves and weights
44. Primary metering jets (2)
45. Secondary metering jets (2)
46. Fuel bowl baffle plates (2)
47. Idle mixture screw and spring (2)
48. By-pass idle air adjustment screw and spring
49. Idle speed screw

PARTS LISTING	WCFB EXPLODED VIEW

1. Pin spring
2. Fast idle rod
3. Throttle connector rod
4. Clevis clip
5. Choke connector rod
6. Dust cover screw (2)
7. Dust cover
8. Dust cover gasket
9. Metering rod (2)
10. Pump connector link
11. Air horn screws (16)
12. Air horn assembly
13. Float lever pin (2)
14. Primary float and lever assembly
15. Secondary float and lever assembly
16. Vacuum piston
17. Primary needle and seat assembly
18. Secondary needle and assembly
19. Air horn gasket
20. Filter nut and *gasket
21. Bowl filter gasket
22. Bowl filter
23. Filter spring
24. Vent arm attaching screw
25. Vent arm
26. Pump operating lever and countershaft
27. Metering rod arm assembly
28. Pump arm assembly
29. Vacuummeter piston link
30. Choke lever assembly
31. Choke valve attaching screw (2)
32. Choke valve
33. Choke shaft and lever assembly
34. Pump plunger assembly
35. Lower pump spring
36. Intake check ball retainer
37. Pump intake check ball
38. Vacuum piston spring
39. Low speed jet (2)- some models
40. Pump jet housing attaching screw
41. Pump jet and housing assembly
42. Pump jet housing gasket
43. Pump discharge check needle
44. Primary metering rod jet (2)
45. Secondary metering jet (2)
46. Body flange attaching screw
47. Body flang e gasket
48. Coil housing attaching screw (3)
49. Coil housing retainer (1 or 3)
50. Thermostatic coil & housing assembly
51. Coil housing gasket
52. Choke baffle plate
53. Choke piston lever, link and pin screw
54. Choke shaft and lever
55. Choke piston link and piston assembly
56. Piston housing attaching screw (3)
57. Piston housing
58. Piston housing gasket
59. Idle adjustment screw and spring (3)
60. Body flange assembly
61. Main body
62. Secondary Control Diaphragm
63. Starter Switch

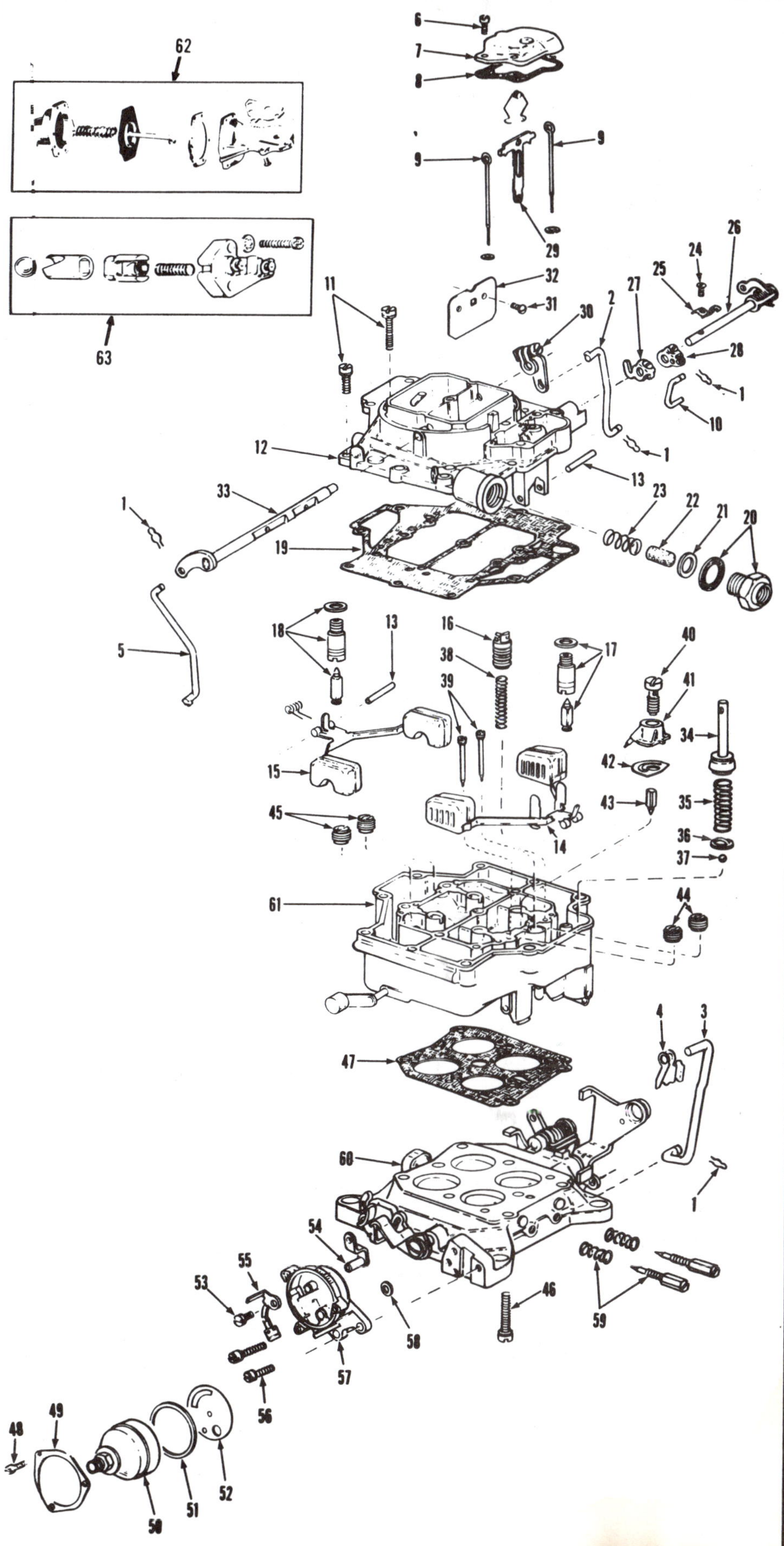